INTEGRITY OF STRUCTURES AND FLUID SYSTEMS, PIPING AND PIPE SUPPORTS, AND PUMPS AND VALVES
– 1997 –

presented at

THE 1997 ASME PRESSURE VESSELS AND PIPING CONFERENCE
ORLANDO, FLORIDA
JULY 27–31, 1997

sponsored by

THE PRESSURE VESSELS AND PIPING DIVISION, ASME

principal editor

HOWARD H. CHUNG
ARGONNE NATIONAL LABORATORY

contributing editors

EVANS C. GOODLING, JR.
PARSONS POWER GROUP, INC.

L. IKE EZEKOYE
WESTINGHOUSE ELECTRIC CO.

THE AMERICAN SOCIETY OF MECHANICAL ENGINEERS
United Engineering Center / 345 East 47th Street / New York, N.Y. 10017

ISBN No. 0-7918-1573-0

Library of Congress Number 97-73606

FOREWORD

This special publication contains nineteen papers presented at the 1997 ASME Pressure Vessels and Piping Conference, July 27-31, 1997 in Orlando, Florida. These sessions were sponsored by the Senate and the Operations, Applications, and Components (OAC) Committee of the ASME Pressure Vessels and Piping Division (PVPD). The presented papers cover a wide variety of topics on structural mechanics of pressure vessels, piping, pipe supports, pressure retaining equipment, pumps, and valves. These papers were presented at the following technical sessions:

- Structural Mechanics of Pressure Vessels and Piping

- Design, Analysis, Maintenance of Pipe Supports, Restraints, and Other Components

- Component Analysis and Evaluations (Pumps and Valves)

- Student Paper Competition Session

Each paper has been subjected to a formal peer review in accordance with the requirements of the ASME. Reviewers' comments have been addressed prior to publication of the final manuscripts.

We would like to thank the authors for their contributions to this volume and the reviewers for their constructive comments to improve the quality of the presented papers. We also wish to express our gratitude to the 1997 PVP Conference Technical Program Chairman, Dr. T. H. Liu of Westinghouse Electric Corporation and the 1997 PVPD Senate President, Mr. Greg L. Hollinger of Babcock & Wilcox who provided strong support and guidance for this publication.

Howard H. Chung
Argonne National Laboratory
Argonne, Illinois

Evans C. Goodling, Jr.
Parsons Power Group, Inc.
Reading, Pennsylvania

L. Ike Ezekoye
Westinghouse Electric Corporation
Pittsburgh, Pennsylvania

CONTENTS

INTEGRITY OF STRUCTURES AND FLUID SYSTEMS

Introduction

Howard H. Chung
Argonne National Laboratory
Argonne, Illinois

Nine papers in this Chapter address a variety of design, fabrication, operation, and inspection issues for pressure vessel, steam generator, piping, snubbers, and other related pressure components.

Two papers by *B. Lin and D. W. Nicholson* concern incremental finite element algorithms to model thermomechanical contact. In Part I, a tangent stiffness matrix was explicitly derived using Kronecker product notation and a gap element based on a nonlinear conductive elastic foundation model was introduced. In Part II, the formulation is applied to elastomeric seals and gaskets which are essential components in pressure vessel systems. A special purpose finite element code implementing the general formulation has been written and applied to a natural rubber seal which is subject to thermal and mechanical loading and confinement.

Pipe tees and lateral connections are essential components in process and power generation facilities. The lateral nozzle makes an elliptical opening on the pipe or vessel surface which has a higher stress concentration than the standard 90o nozzle and causes major safety concerns especially in nuclear power piping design. The paper by *J. J. Xu, B. C. Sun, and B. Koplik* presents a comprehensive analysis conducted to investigate the local pressure stress around a pipe-nozzle with a 45-degree angle of intersection.

In a series of two papers, *G. B. Stephenson, G. H. Lindsey, Y. W. Kwon, and K. S. Song* investigated inelastic stress concentrations around notches. In Part I, a series of numerical studies were conducted to evaluate the Glinka and Neuber expressions to predict inelastic stress concentrations around notches of specimens under uniaxial tensile loads. They investigated both plane stress and plane strain conditions. In Part II, experimental studies were conducted to further evaluate Neuber's and Glinka's predictions for different loadings and materials.

Fatigue life prediction is a subject of great interest for gas turbine components where high-strength steels and ceramics are used increasingly to achieve enhanced durability under severe operational stresses and temperatures. Two papers by *D. W. Nicholson and Peizhong Ni* address the application probabilistic fracture mechanics (PFM) models to gas turbine structures. In the first paper, a probabilistic model was developed which combines rigorous fracture criteria with extreme value methods to describe the strength of brittle elastic plates containing pre-existing cracks with random number, length, and orientation . Existing mixed-mode fracture criteria have been extended to accommodate biaxial loads plus torsion. In the second paper, they introduces a probabilistic model based on established criteria from fracture mechanics where crack number, length, and orientation at the onset of fatigue are treated as random variables. These models accommodate multi-axial loads and can be useful for analyzing gas turbine engine components.

The paper by *B. Blyukher* summarizes the basic safety requirements of compressor systems commonly used in process industries, R&D organizations, and test facilities. In developing these requirements, the author has consulted with OSHA, ANSI, and other industry standards and regulations. Furthermore, the author presents some specific recommendations for preventing compressor accidents and promoting safe operation.

The last paper by *M. Moatamedi, B. C. R. Ewan, and J. L. Wedring* is the winning entry to the student paper competition at the 1997 Pressure Vessels and Piping Conference. This paper presents a transient analysis of a cylindrical vessel under asymmetric external as well as internal impulsive loading carried out using a finite element method. The authors focused their investigations on the effects of explosion duration and the natural frequency of the structure on the response of the structure. It was found out that by increasing the duration, both the maximum displacement and the time in which the maximum displacement occurs increase.

PVP-Vol. 356, Integrity of Structures and Fluid Systems,
Piping and Pipe Supports, and Pumps and Valves
ASME 1997

INCREMENTAL FINITE ELEMENT METHOD FOR THERMOMECHANICAL CONTACT
PART I : TANGENT STIFFNESS MATRIX

Baojiu Lin, Ph. D.
Postdoctoral Scientist

David W. Nicholson, Ph. D.
Professor and Director

Institute for Computational Engineering
Department of Mechanical, Material, and Aerospace Engineering
University of Central Florida
Orlando, FL 32816

ABSTRACT

This investigation concerns the development of incremental finite element algorithms to support design improvements and thermomechanical analysis of components subject to variable thermomechanical contact. The basic quantity which arises in the formulation is a *tangent stiffness matrix*, which in essence serves as a Jacobian matrix for the purpose of solution by Newton iteration. This quantity is derived explicitly in Part I in compact form using Kronecker product notation. Existing finite element codes with contact capabilities are based on approximations which are suitable for small geometric change, and in any event do not couple thermal and mechanical effects in contact. Here, a gap element suitable for modeling variable contact is introduced based a nonlinear conductive elastic foundation model. It has no small deformation restrictions and avoids some computational difficulties posed by commonly used bilinear elastic models. The gap function is used to derive contributions to the Jacobian matrix from the contact condition. Owing to the power of Kronecker product notation, compact expressions are derived for several otherwise intractable quantities arising in the Jacobian matrix, such as the derivative of the surface normal vector with respect to the displacement vector. In Part II, the formulation is applied to elastomeric seals and gaskets, which of course are essential components in pressure vessel systems.

INTRODUCTION

Themomechanical contact is important in many applications, such as seals and gaskets in pressure vessels, where they are confined while being subjected to high temperature and pressure. Components with good contact behavior will enhance the performance such systems. Design based on trial and error is slow and expensive. Finite element simulation is one of the powerful tools supporting computer-aided design. Nonlinear finite element analysis codes are expected by engineers to assist them in design, manufacture, and installation of components. Unfortunately, existing finite element codes familiar to the authors do not accommodate variable thermomechanical contact. Recently, a systematic investigation has been reported by Nicholson and Lin (1996b, 1996c, 1996d) and Lin (1996). It included a new model for variable thermomechanical contact using a special type of nonlinear elastic foundation involving a gap function. In many analyses a Fully Lagrangian formulation is adopted in which the traction vector is assumed to be prescribed on the undeformed (external) surface. The incremental equations may then be formulated as a realization of Newton iteration, with the *tangent stiffness matrix* serving as a *Jacobian matrix*. In many applications, the deformed configuration is distinct from the undeformed configuration. In the current investigation, the Fully Lagrangian formulation is likewise adopted. However, in contrast to previous investigations, the traction and heat flux vectors are expressed in terms of the *deformed* coordinates, Fourier's law of heat conduction is expressed in terms of the *deformed* coordinates, and variable *thermomechanical* contact is modeled using the deformed coordinates. In the following sections, detailed expressions for the ensuing finite element equation and its Jacobian matrix are presented in compact form using Kronecker product and other notations.

MATHEMATICAL BACKGROUND AND NOTATION

The current investigation concerns a body whose undeformed configuration has coordinates X, with uniform temperature T_0. The deformed coordinates are denoted by $x(X, t)$, and the temperature is $T(X,t)$. Kronecker tensor notation is used throughout (cf. Graham, 1981); it permits compact expressions for elaborate quantities such as the *tangent modulus matrix* for hyperelastic materials (Nicholson, 1995). If $\mathbf{\varepsilon}$ denotes the Lagrangian strain, its vectorial counterpart e is given by

$$
\begin{aligned}
\mathbf{e} &= VEC(\mathbf{\varepsilon}) \\
&= \{e_{11}\ e_{21}\ e_{31}\ e_{21}\ e_{22} \\
&\quad e_{32}\ e_{13}\ e_{23}\ e_{33}\}^T.
\end{aligned}
\tag{1}
$$

The trace of a product of two matrices is

$$
\begin{aligned}
tr(\mathbf{AB}) &= VEC^T(\mathbf{A}^T)\,VEC(\mathbf{B}) = \mathbf{a}^T\mathbf{b}, \\
\mathbf{a} &= VEC(\mathbf{A}^T), \qquad \mathbf{b} = VEC(\mathbf{B}).
\end{aligned}
\tag{2}
$$

The Kronecker products of two vectors and of two matrices are

$$
\mathbf{a} \otimes \mathbf{b} = VEC(\mathbf{ba}^T) \qquad \mathbf{A} \otimes \mathbf{B} = \begin{bmatrix} a_{11}\mathbf{B} & a_{12}\mathbf{B} & \cdots \\ a_{21}\mathbf{B} & \cdots & \cdots \\ \cdots & \cdots & \cdots \end{bmatrix}.
\tag{3}
$$

Also,

$$
VEC(\mathbf{AB}) = \mathbf{I} \otimes \mathbf{A}\, VEC(\mathbf{B}) = \mathbf{B}^T \otimes \mathbf{I}\, VEC(\mathbf{A}).
\tag{4}
$$

With repeated indices implying summation, the following relations will prove useful:

$$
\mathbf{a}^T\mathbf{Bc} = \mathbf{c}^T \otimes \mathbf{a}^T\, VEC(\mathbf{B}).
\tag{5}
$$

The primary quantities requiring interpolation models are the displacement u, the Lagrangian strain $e = VEC(\mathbf{\varepsilon})$, the Cauchy-Green strain $c = VEC(C)$, the temperature T, the temperature gradient $\nabla_0 T$, and the temperature-adjusted pressure π, defined below. The interpolation models are:

$$
\mathbf{u} = \mathbf{N}^T(\mathbf{X})\,\mathbf{\gamma}(t) \qquad VEC(\mathbf{F}-\mathbf{I}) = \mathbf{M}_1^T\mathbf{\gamma}
$$

$$
VEC(\mathbf{F}^T-\mathbf{I}) = \mathbf{M}_2^T\mathbf{\gamma} \qquad \delta\mathbf{e} = \mathbf{\beta}_1^T\mathbf{\gamma}
$$

$$
\delta\mathbf{c} = 2\delta\mathbf{e} = 2\mathbf{\beta}_1^T\mathbf{\gamma} \qquad \mathbf{\beta}_1 = \frac{1}{2}[\mathbf{I} \otimes \mathbf{F}^T\mathbf{M}_1^T + \mathbf{F}^T \otimes \mathbf{I}\mathbf{M}_2^T]
\tag{6}
$$

$$
T = \mathbf{v}^T(\mathbf{X})\,\mathbf{\theta}(t) \qquad \nabla_0 T = \mathbf{\beta}_T^T(\mathbf{X})\,\mathbf{\theta}
$$

$$
\pi = \mathbf{\zeta}^T(\mathbf{X})\,\mathbf{\psi}(t).
$$

The matrices N, v, and ς are *shape functions*, the subscript "$_0$" in ∇_0

refers to the *undeformed coordinates*, $\pi = p/f^3(T)$ where $f(T) = 1/[1 + \alpha(T-T_0)/3]$, α is the volume coefficient of thermal expansion, and p is the true hydrostatic pressure. Finally, γ, θ, and ψ are *global* vectors of the nodal values of u, T, and π, respectively. The invariants of Cauchy-Green deformation tensor C are denoted by I_1, I_2, and I_3. The following expressions are quoted from Nicholson (1995):

$$
\mathbf{n}_i^T = \frac{\partial I_i}{\partial \mathbf{c}} \qquad \mathbf{A}_i = \frac{\partial \mathbf{n}_i}{\partial \mathbf{c}}
$$

$$
\mathbf{c}_2 = VEC(\mathbf{C}^2) \quad \mathbf{n}_1 = VEC(\mathbf{I}) = \mathbf{1}
$$

$$
\mathbf{n}_2 = I_1\mathbf{1} - \mathbf{c} \quad \mathbf{n}_3 = I_2\mathbf{1} - I_1\mathbf{c} + \mathbf{c}_2
\tag{7}
$$

$$
\mathbf{A}_1 = \mathbf{0} \qquad \mathbf{A}_2 = \mathbf{1}\mathbf{1}^T - \mathbf{I}_9
$$

$$
\mathbf{A}_3 = \mathbf{I} \otimes \mathbf{C} + \mathbf{C} \otimes \mathbf{I} - \mathbf{1}\mathbf{c}^T + \mathbf{c}\mathbf{1}^T) + I_1(\mathbf{1}\mathbf{1}^T - \mathbf{I}_9).
$$

Also

$$
\frac{dJ}{d\mathbf{c}} = \frac{1}{2J}\mathbf{n}_3^T \qquad J = \sqrt{I_3}.
\tag{8}
$$

and the Cayley-Hamilton theorem is readily shown to imply that

$$
VEC(\mathbf{C}^{-1}) = I_3^{-1}\mathbf{n}_3.
\tag{9}
$$

VARIABLE THERMOMECHANICAL CONTACT

A general expression of boundary traction and heat flux was introduced by the authors (Nicholson and Lin, 1996d) in the following form

$$
\begin{cases} d\mathbf{t} = \underline{d\mathbf{t}} + \mathbf{A}_{MM}^T d\mathbf{\gamma} + \mathbf{a}_{MT} dT \\ d(\mathbf{n}^T\mathbf{q}) = \underline{d(\mathbf{n}^T\mathbf{q})} + \mathbf{a}_{TM}^T d\mathbf{\gamma} + a_{TT} dT \end{cases}
\tag{10}
$$

where $\underline{d\mathbf{t}}$ and $\underline{d(\mathbf{n}^T\mathbf{q})}$ are prescribed. A_{MM}, a_{MT}, a_{TM}, and a_{TT} are defined below for variable thermomechanical contact.

A. Mechanical Contact

In commercial finite element codes such as ANSYS(1995) and COSMOS/M(1990), gap elements are used to treat contact. They can be interpreted as nonlinear elastic foundations. In this investigation, the gap function $g(x)$ is defined by

$$
\begin{cases} \mathbf{y} = \mathbf{x} + g(\mathbf{x})\mathbf{n}(\mathbf{x}) & \mathbf{x}\ on\ S_c \\ 0 = \psi_r(\mathbf{y}) & \mathbf{y}\ on\ S_r \end{cases}
\tag{11}
$$

where S_c and S_r are the candidate contact surfaces of the contactor and the rigid foundation, respectively. ψ_r is the algebraic equation for S_r. $\mathbf{x}$ and $\mathbf{y}$ are the positions of possible contact points on S_c and S_r, respectively. $\mathbf{n}$ is the unit vector normal to the surface S_c.

It is assumed that the mechanical contact between a contactor and a target may be accurately simulated through the following special nonlinear elastic foundation model. Namely, the increment of the contact traction is assumed as an explicit function of the gap as follows,

$$\Delta t_c^{(n)} = -K(G) \Delta u_{cn}^{(n)}$$
$$G = g^{(n)} \mathbf{n}^{(n)T} \mathbf{n}^{(n)} - \mathbf{n}^{(n)T} \Delta \mathbf{u}^{(n)} . \quad (12)$$

Here the superscript "(n)" means the values at the beginning of the $(n+1)$st load step, and G is an updated gap function for the $(n+1)$st step and $\underline{\mathbf{n}}$ is a unit vector normal to the target (cf. Lin (1996)).

In Lin (1996), we introduced a continuous model for the elastic foundation:

$$K(G) = \frac{k_H}{\pi} \left\{ \frac{\pi}{2} - \tan^{-1}[\alpha_k(G-e_r)] \right\} + k_L . \quad (13)$$

Here, k_H is a stiffness value much larger than the characteristic stiffness of the adjacent contactor elements. It approximates the rigidity of the foundation. But, in practice, its stiffness is restricted by the convergence requirements of iteration. Consequently, it is determined by numerical experiment. k_L is a stiffness value lower than that of the adjacent contactor elements. It is used to avoid possible rigid body motion in finite element computation. α_k is a large number which determines the transition range from the low stiffness to high stiffness. e_r is a very small number which is used to adjust the initial contact stiffness. To avoid jumps from low stiffness to high stiffness during iteration, the size of the load step is selected such that the increment of displacement in each load step is in the transition range. This continuous function is shown graphically in Fig.(1). This continuous model contrasts with, and we believe avoids disadvantages of, bilinear models used in a numbers of finite element codes.

Now we proceed to determine $\underline{dt}$, A_{MM}, and a_{MT} for the case of variable mechanical contact. The contact traction is obtained as

$$\begin{aligned} \mathbf{t}_c &= (t_n^{(n)} + \Delta u_n^{(n)}) \mathbf{n} \\ &= [t_n^{(n)} + K(G)(\mathbf{u}^T \mathbf{n} - \mathbf{u}^{(n)T} \mathbf{n}^{(n)})] \mathbf{n} . \end{aligned} \quad (14)$$

Then

$$\begin{aligned} d\mathbf{t}_c &= (t_n^{(n)} + K(G)\Delta u_n) \frac{\partial \mathbf{n}}{\partial \boldsymbol{\gamma}} d\boldsymbol{\gamma} \\ &\quad + \mathbf{n} \mathbf{u}^T \mathbf{n} \frac{\partial K(G)}{\partial G} \frac{\partial G}{\partial \boldsymbol{\gamma}} d\boldsymbol{\gamma} \\ &\quad + \mathbf{n} K(G) \mathbf{n}^T \mathbf{N}^T d\boldsymbol{\gamma} \\ &\quad + \mathbf{n} K(G) \mathbf{u}^T \frac{\partial \mathbf{n}}{\partial \boldsymbol{\gamma}} d\boldsymbol{\gamma} \\ &= \underline{dt} + A_{MM}^T d\boldsymbol{\gamma} + \mathbf{a}_{MT} dT , \end{aligned} \quad (15)$$

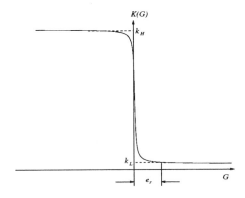

Figure 1 Continuous model for elastic foundation

where

$$\underline{dt} = 0 \qquad \mathbf{a}_{MT} = 0$$
$$\begin{aligned} A_{MM}^T &= (t_n^{(n)} + K(G)\Delta u_n) \frac{\partial \mathbf{n}}{\partial \boldsymbol{\gamma}} \\ &\quad + \mathbf{n} \mathbf{u}^T \mathbf{n} \frac{\partial K(G)}{\partial G} \frac{\partial G}{\partial \boldsymbol{\gamma}} \\ &\quad + \mathbf{n} K(G) \mathbf{n}^T \mathbf{N}^T \\ &\quad + \mathbf{n} K(G) \mathbf{u}^T \frac{\partial \mathbf{n}}{\partial \boldsymbol{\gamma}} . \end{aligned} \quad (16)$$

The following quantities are elaborate and are derived explicitly in Lin (1996).

$$\frac{\partial \mathbf{n}}{\partial \boldsymbol{\gamma}} = \left(\frac{\mathbf{n} \mathbf{n}_3^T}{I_3} - \frac{2 \mathbf{n} \mathbf{w}^T}{\mu} \right) \boldsymbol{\beta}_1^T - \mathbf{n}^T \otimes \mathbf{F}^{-T} \mathbf{M}_1 \quad (17)$$

$$\mu = \sqrt{\mathbf{n}_0^T \otimes \mathbf{n}_0^T \mathbf{n}_3} \quad (18)$$

$$\frac{\partial G}{\partial \boldsymbol{\gamma}} = -\mathbf{n}^T \mathbf{N} \quad (19)$$

$$\mathbf{w}^T = \frac{\mathbf{n}_0^T \otimes \mathbf{n}_0^T \mathbf{A}_3}{\mu} \quad (20)$$

$$\mathbf{F}^{-T} = \frac{\mathbf{F}}{I_3} (\mathbf{C}2 - I_1 \mathbf{C} + I_2 \mathbf{I}) \quad (21)$$

where $\mathbf{n}$ and $\mathbf{n}_0$ are the unit vectors normal to deformed and undeformed surfaces, respectively. They are related by.

$$n = \frac{JF^{-T}n_0}{\mu}.$$ (22)

Finally, we note that no simplifications have been made. This is in contrast to these formulations in which approximations are commonly made which assume small deformation.

B. Variable Thermal Contact

The interface formed by contactor and target affects resistance to heat flow between the contacting materials. The heat flow through the thermal contact surfaces can be expressed as

$$n^T q = \alpha_h(G)(T_r - T)$$ (23)

where T_r is the foundation temperature. Here the heat conductance α_h is assumed as a function of gap G. It must be determined experimentally.

Now we proceed to identify $d(n^T q)$, a_{TM}, and a_{TT} for the case of variable thermal contact from

$$d(n^T q) = \frac{\partial \alpha_h}{\partial G}\frac{\partial G}{\partial \gamma}(T_r - T)d\gamma + \alpha_h dT$$
$$= d(n^T q) + a_{TM}^T d\gamma + a_{TT} dT.$$ (24)

Hence

$$d(n^T q) = 0 \qquad a_{TT} = -\alpha_h$$
$$a_{MT} = \frac{\partial \alpha_h}{\partial G}\frac{\partial G}{\partial \gamma}(T_r - T).$$ (25)

GOVERNING EQUATIONS OF MIXED FIELDS

A. Mechanical Equilibrium

Mechanical equilibrium is expressed by the Principle of Virtual Work:

$$\int \delta e^T s dV_0 = \int \delta u^T t dS_0$$ (26)

in which V_0 and S_0 are the volume and surface area in the undeformed configuration.

B. Thermal Equilibrium

We now present equations for the thermal field, following Chung (1988). Neglecting kinetic energy and heat sources, and assuming only conductive heat transfer, the equation of thermal equilibrium is written

in local form as

$$\eta = -\frac{\nabla^T q}{\rho_0 T}$$ (27)

where the Fourier's law for heat conduction is expressed in the *deformed coordinates*, as follows,

$$q = -k(T)\nabla T, \quad k(T) \geq 0.$$ (28)

To accommodate materials such as elastomers η now is regarded as a function of e, T, and π (cf. Nichoslon and Lin (1996a, 1996c)). From Eq.(27), follows that

$$-\frac{\nabla^T q}{\rho T} = \frac{\partial \eta}{\partial T}\dot{T} + \frac{\partial \eta}{\partial e}\dot{e} + \frac{\partial \eta}{\partial \pi}\dot{\pi}$$ (29)

Unfortunately, the thermal field presents difficulties since it is expressed in terms of rates. The solutions for all u, T, and π are assumed known at time $t_n = nh$, where $n = 1, 2, 3, ...,$ and h is the time step. The solution is sought at $t_{n+1} = (n+1)h$. However, Eq.(27) is replaced using the trapezoidal rule

$$\rho_0 \eta = -\frac{h}{2}\frac{\nabla^T q}{T} + \rho_0 \eta_n - \frac{h}{2}\frac{\nabla^T q}{T}|_n + o(h^2).$$ (30)

Here," $|_n$" implies evaluation at t_n. However, $|_{n+1}$ corresponding to the current time step, is not displayed. Neglecting higher order terms in h, the variational equation for the thermal field is now written as

$$\int \delta T \rho_0 \eta dV_0 = \int \delta T\left(-\frac{h}{2}\frac{\nabla^T q}{T} + \rho_0 \eta_n - \frac{h}{2}\frac{\nabla^T q}{T}|_n\right)dV.$$ (31)

C. Near-incompressibility Constraint

The constraint equation for π is:

$$\int \delta \pi \rho_0 \omega dV_0 = 0$$ (32)

where

$$\omega = \frac{\pi}{\rho_0 \kappa} + \frac{f^3 J - 1}{\rho_0}.$$ (33)

This equation applies to elastomers but can be eliminated in analyses involving materials which are compressible.

INCREMENTAL FINITE ELEMENT EQUATIONS

After some manipulation, the final finite element equations for the mixed fields are written in the form

$$r = f(t, \theta(t)) \qquad (34)$$

where

$$r = \begin{Bmatrix} r_M \\ r_\pi \\ r_T \end{Bmatrix} \quad v = \begin{Bmatrix} \gamma \\ \phi \\ \theta \end{Bmatrix} \quad f = \begin{Bmatrix} f_M \\ 0 \\ f_T \end{Bmatrix} \qquad (35)$$

$$r_M = \int \beta s \, dV_0 \qquad (36)$$

$$f_M = \int N t \mu \, dS_0 \qquad (37)$$

$$r_\pi = \int \zeta \rho_0 \omega \, dV_0 \qquad (38)$$

$$r_T = \int \mathbf{v} \rho_0 \eta \, dV_0 \qquad (39)$$

$$
\begin{aligned}
f_T = & \int \mathbf{v} \rho_0 \eta_n \, dV_0 - \frac{h}{2} \int \left[\frac{\mathbf{v}(n^T q) \mu}{T} \right]_n dS_0 \\
& + \frac{h}{2} \int \left[\frac{k \beta_T (J C^{-1}) \nabla_0 T}{T} \right]_n dV_0 \\
& - \frac{h}{2} \int \frac{\mathbf{v}(n^T q) \mu}{T} \, dS_0 \\
& + \frac{h}{2} \int \frac{k \beta_T (J C^{-1}) \nabla_0 T}{T} \, dV_0
\end{aligned} \qquad (40)
$$

and the incremental finite element equations can be expressed as

$$\delta v^T [r(v,t) - g_0(v,t) - h g_1(v,t)] = 0 \qquad (41)$$

with

$$g_0^T(v,t) = \{ g_M^T \ 0^T \ g_{0T}^T \}^T \qquad (42)$$

$$g_1^T(v,t) = \{ 0^T \ 0^T \ g_{1T}^T \}^T \qquad (43)$$

$$g_M = \int N t \mu \, dS_0 \qquad (44)$$

$$g_{0T} = \int \mathbf{v} \rho_0 \eta \big|_n dV_0 - \frac{h}{2} \mathbf{v} \frac{\nabla^T q J}{T} \big|_n dV_0 \qquad (45)$$

$$g_{1T} = -\frac{1}{2} \int \mathbf{v} \frac{\nabla^T q J}{T} \, dV_0 \qquad (46)$$

The Jacobian matrix J now incoporates several *boundary* terms into the tangent stiffness matrix. The primary object of the present investigation is to introduce compact expressions for the boundary terms. Kronecker product notation permits compact expressions and is used throughout (Graham, 1981, Nicholson, 1995). The Jacobian matrix J is used in solving the incremental equation

$$J dv = dg_M + h dg_{1T} \qquad (47)$$

$$J = (K + hR) + (B_t + h B_q) + (B_{ct} + h B_{cq}) \qquad (48)$$

$$dg_M = \int N dt \mu \, dS_0 \qquad (49)$$

$$dg_{1T} = -\frac{1}{2} \int \frac{\mathbf{v} d(n^T q)}{T} \mu \, dS_0 \qquad (50)$$

The details of the matrices K, R, B_t, and B_q are reported in Nicholson and Lin (1996c, 1996d), and Lin (1996). B_{ct} and B_{cq} are obtained as

$$\begin{bmatrix} B_{CMM} & 0 & B_{CMT} \\ 0^T & 0 & 0 \\ 0^T & 0^T & 0 \end{bmatrix} \quad B_{cq} = \begin{bmatrix} 0 & 0 & 0 \\ 0^T & 0 & 0 \\ B_{cTM}^T & 0^T & B_{cTT} \end{bmatrix} \qquad (51)$$

$$B_{CMM} = \int N a_{MM}^T \mu \, dS_{c_0} \qquad (52)$$

$$B_{CMT} = \int N a_{MT} \mathbf{v}^T \mu \, dS_{c_0} \qquad (53)$$

$$B_{cTM}^T = -\frac{1}{2} \int \frac{\mu \mathbf{v} a_{TM}^T}{T} \, dS_{c_0} \qquad (54)$$

$$B_{cTT} = -\frac{1}{2} \int \frac{\mu a_{TT} \mathbf{v} \mathbf{v}^T}{T} \, dS_{c_0}. \qquad (55)$$

Computational results of the derived formulation are presented in Part II, which concerns an elastomeric seal which is submitted to force and heat, and is pressed into a rigid well.

DISCUSSION AND CONCLUSIONS

A previous formulation for coupled thermomechanical response by finite elements has been extended to the case in which there is variable thermomechanical contact. In the current investigation, traction and heat flux are assumed to be prescribed on the deformed surface and Fourier's law for heat conduction is expressed in terms of the deformed coordinates. A contact model is formulated which avoids simplifications which assume small deformation. Numerical experiments show the continuously nonlinear elastic foundation model used in current investigation has advantages over the conventional bilinear model which is used widely in commercial finite element codes.

REFERENCES

ANSYS User's Manual, Ver. 5, 1993, Swanson Associates.

COSMOS/M User's Guide Vol. I, Ver. 1.6, 1990, SRAC., California,

Chung, T. J., 1988, "Continuum Mechanics" Prentice Hall, Englewood Cliffs, New Jersey.

Chandrasekharaiah, D. S. and Debnath,I.., 1994, "Continuum Mechanics", Academic Press, New York.

Graham, A., 1981, "Kronecker Products and Matrix Calculus with Applications", Ellis Horwood Ltd., Chichester, U. K..

Lin, Baojiu, 1996, "Finite Element Methods for Thermohyperelastic Bodies under Non-classical Boundary Conditions", Ph. D. dissertation, University of Central Florida, Orlando.

Nicholson, D. W., 1995, "Tangent Modulus Matrix for Finite Element Analysis of Hyperelastic Materials", ACTA MECHANICA, Vol. 112, pp. 187-201.

Nicholson, D. W. and Lin, B., 1996a, "Thermohyperelastic Constitutive Model for Near-Incompressible elastomers", ACTA MECHANICA, Vol. 116, pp. 15-28.

Nicholson, D. W. and Lin, B., 1996b, "Finite Element Method for Thermomechanical Response of Near-incompressible Elastomers", ACTA MECHANICA, in press.

Nicholson, D. W. and Lin, B., 1996c,"Incremental Finite Element Equations for Thermomechanical Response of Elastomers: Effect of Boundary Conditions Including Contact", ACTA MECHANICA, in press .

Nicholson, D. W., Nelson, N. W., Lin, B., and Farinella, A.,1996d, "Finite Element Analysis of Hyperelastic Components", Applied Mechanics Reviews, in review.

PVP-Vol. 356, Integrity of Structures and Fluid Systems,
Piping and Pipe Supports, and Pumps and Valves
ASME 1997

INCREMENTAL FINITE ELEMENT METHOD FOR THERMOMECHANICAL CONTACT
PART II: APPLICATION TO ELASTOMERIC SEALS

Baojiu Lin, Ph. D.
Postdoctoral Scientist

David W. Nicholson, Ph. D.
Professor and Director

Institute for Computational Engineering
Department of Mechanical, Material, and Aerospace Engineering
University of Central Florida
Orlando, FL 32816

ABSTRACT

In Part I, an incremental finite element formulation was introduced to model thermomechanical contact, which is a type of boundary condition which occurs widely in pressure vessel applications, for example at supports. A new gap function was introduced and a compact expression was derived for the contribution of contact to the *tangent stiffness matrix*, which serves as the Jacobian matrix for solution by Newton iteration. Part II applies the formulation to elastomeric seals, which are essential components of pressure vessel systems. Of particular concern is the accuracy with which high pressures ensuing from confinement are modeled. A three field formulation is adopted, combing a displacement field, a temperature field, and a pressure field introduced to satisfy the constraint of near-incompressibility. Typically, commercial finite element codes model elastomers using a hyperelastic element, which is not coupled to the thermal field. Here, a thermohyperelastic constitutive model for near-incompressible elastomers, introduced by the authors, is used. The *tangent modulus matrix* ensuing from the constitutive model is derived in compact form using Kronecker product notation. An application of great interest is seals. A special purpose finite element code implementing the general formulation has been written and applied to a natural rubber seal which is subject to thermal and mechanical loading and confinement., say by being pressed into a well. The computations have been validated in various ways, and are illustrated in graphical form. In particular, pressure contours in the seal are shown as a function of degree of compression and of time. The great amplification of pressure due to confinement is captured.

INTRODUCTION

Elastomeric components are important in many applications, such as seals and gaskets in pressure vessels, where may be confined while being subjected to high temperature and pressure. High performance elastomeric components will enhance the performance of such systems. Design based on trial and error is slow and expensive. Finite element simulation is one of the powerful tools supporting computer-aided design. Nonlinear finite element analysis codes are expected by engineers to assist them in design, manufacture, and installation of elastomeric components. Unfortunately, existing finite element codes familiar to the authors do not accommodate coupled thermomechanical response of near-incompressible bodies with variable thermomechanical contact. Recently, a systematic investigation has been reported by Nicholson and Lin (1996a, 1996b, 1996c, 1996d), and Lin (1996). It included a new thermohyperelastic constitutive model for near-incompressible elastomers, a new model for variable thermomechanical contact using a special type of nonlinear elastic foundation involving a gap function, and a mixed three-field (the displacement field γ, the pressure field ψ, and the temperature field θ) nonlinear finite element algorithm which accommodates large deformations and variable thermomechanical contact. Here, the formulation is summarized in Part I and the contribution of contact to the tangent stiffness matrix is presented in detail. The formulation is applied to simulation of an elastomeric o-ring seal submitted to force and heat and pressed into a rigid well.. A special purpose finite element code has been written to perform the simulation.

THERMOMECHANICAL CONSTITUTIVE MODEL FOR ELASTOMERS

In Nicholson and Lin (1996a), a thermohyperelastic constitutive model was introduced for near-incompressible elastomers, expressed by

$$\phi = \phi_M (I_1 I_3^{-1/3}, \ I_2 I_3^{-2/3}) + c_e T \left[1 - \ln\left(\frac{T}{T_o}\right) \right] - \frac{\pi}{\rho_o} (f^3(T) J - 1) - \frac{\pi^2}{2\kappa\rho_o} + \phi_o \quad (1)$$

where ρ_o is the density, κ is the bulk modulus, and c_e is related to the specific heat of uniform strain, and ϕ_o is a constant.

Now, if we take π as varying independently, then the stationary conditions of the expression (1) become

$$\sigma = \rho_o \frac{\partial \phi_M}{\partial \varepsilon} + \kappa (f^3(T) J - 1) f^3(T) J C^{-1} \quad (2)$$

$$\eta = \alpha\kappa\rho_o (f^3(T) J - 1) J f^4(T) + c_e \ln\left(\frac{T}{T_o}\right) \quad (3)$$

$$0 = \frac{\pi}{\kappa} + (f^3(T) J - 1) \quad (4)$$

where Eq.(4) expresses the near-incompressibility constraint. Here σ is the 2nd Piola-Kirchhoff stress tensor and η is the entropy density. The thermohyperelastic constitutive model for incompressible elastomers is a special case of the expression (1), namely, the case when $\kappa \to \infty$.

For later use, this constitutive model is specialized to provide a thermohyperelastic near-incompressible counterpart of the conventional Mooney-Rivlin elastomer as follows:

$$\phi = C_1 (I_1 I_3^{-1/3} - 3) + C_2 (I_2 I_3^{-2/3} - 3)$$

$$+ c_e T \left[1 - \ln\left(\frac{T}{T_o}\right) \right] - \frac{\pi}{\rho_o} (f^3(T) J - 1) \quad (5)$$

$$- \frac{\pi^2}{2\kappa\rho_o} + \phi_o.$$

DOMAIN CONTRIBUTION TO THERMOHYPERELASTIC TANGENT STIFFNESS MATRIX

Kronecker notation is used as detailed in Part I. Let σ, ε, and C denote the 2nd Piola-Kirchhoff stress, the Lagrangian strain, and the Cauchy-Green strain, and let $s=\text{VEC}(\sigma)$, $e=\text{VEC}(\varepsilon)$, and $c=\text{VEC}(C)$. Interpolation models are introduced as follows for the displacement, temperature, and pressure (for near-incompressibility) fields.

$$u = N^T(X) \gamma(t)$$
$$T = v^T(X) \theta(t) \quad (6)$$
$$\pi = \zeta^T(X) \phi(t)$$

Here γ, θ, and ϕ are vectors of nodal values while N, v, and ζ are shape functions. Also π is an adjusted pressure given by

$$\pi = p / f^3(T) \quad (7)$$

where p is the pressure (one third the trace of the Cauchy stress). The ensuing relations for strain and the temperature gradient are

$$\text{VEC}(F_u) = M_1^T \gamma \qquad \text{VEC}(F_u^T) = M_2^T \gamma$$

$$F_u = F - I \qquad \delta e = \beta_1^T \gamma \quad (8)$$

$$\beta_1 = \frac{1}{2} [I \otimes F^T M_1^T + F^T \otimes I M_2^T] \quad \delta(de^T) s = \delta\gamma^T D_{NL} d\gamma$$

$$D_{NL} = M_1 \sigma \otimes I M_1^T \qquad \nabla_0 T = \beta_T^T(X) \theta$$

where F is the deformation gradient tensor.

The constitutive relations furnish

$$s^T = 2 \sum_i \phi_i n_i^T - \frac{\pi f^3 n_3^T}{J} \quad (9)$$

$$\eta = -\frac{\partial \phi_M}{\partial T} - \frac{\alpha \pi J f^4}{\rho} \quad (10)$$

The specific heat c_e' at constant strain is given by

$$c_e' = T \frac{\partial \eta}{\partial T}\Big|_{e,\pi} = c_e + \frac{4}{3\rho} \alpha^2 \pi J f^5 T \quad (11)$$

The tangent modulus matrix is given by

$$D_{T\pi} = \frac{\partial s}{\partial e}\Big|_{T,\pi} = 4 \sum_i \phi_i A_i + 4 \sum_i \sum_j \phi_{ij} n_i n_j^T - \frac{2\pi f^3}{J} A_3 \quad (12)$$

The relations governing the three fields lead to a system of algebraic equations for which it is attractive to use Newton iteration for numerical solution. Newton iteration requires the Jacobian matrix. Omitting the details, for a vansihing time step, the domain contribution to the Jacobian matrix is given by K, which satisfies

$$v = \begin{Bmatrix} \gamma \\ \psi \\ \theta \end{Bmatrix} \qquad w = \begin{Bmatrix} e \\ \pi \\ T \end{Bmatrix} \qquad (13)$$

$$\delta v^T K v = \int \delta w^T H dw dV \qquad (14)$$

where

$$H = \begin{bmatrix} \dfrac{\partial s}{\partial e}\big|_{\pi,T} & \dfrac{\partial s}{\partial \pi}\big|_{e,T} & \dfrac{\partial s}{\partial T}\big|_{e,\pi} \\[6pt] \rho\dfrac{\partial \omega}{\partial e}\big|_{\pi,T} & \rho\dfrac{\partial \omega}{\partial \pi}\big|_{e,T} & \dfrac{\partial \omega}{\partial T}\big|_{e,\pi} \\[6pt] \rho\dfrac{\partial \eta}{\partial e}\big|_{\pi,T} & \rho\dfrac{\partial \eta}{\partial \pi}\big|_{e,T} & \rho\dfrac{\partial \eta}{\partial T}\big|_{e,\pi} \end{bmatrix}$$

$$\qquad (15)$$

$$= \begin{bmatrix} D_{\pi T} & \dfrac{-f^3 n_3}{J} & \dfrac{\alpha\pi f^4 n_3}{J} \\[6pt] \dfrac{f^3 n_3^T}{J} & \dfrac{1}{\kappa} & -\alpha J f^4 \\[6pt] \dfrac{-\alpha f^4 n_3^T}{J} & -\alpha J f^4 & \rho\dfrac{c_e'}{T} \end{bmatrix}$$

APPLICATION TO ANALYSIS OF CONFINED ELASTOMERIC O-RING SEALS

A special purpose finite element code based on the previous section and the formulation in Part I has been developed to analyze confined elastomeric seals subjected to thermal and mechanical loads. A computational example is presented here for illustration. In this example, an elastomeric o-ring seal with cross-section diameter $d_i=0.5$(cm) is confined in a metallic well as shown in Fig. (1). It is treated as a plane strain problem.

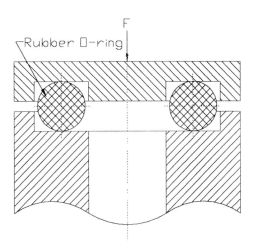

Figure 1 A rubber seal is pressed into a metallic well.

Due to symmetry of the rubber o-ring seal, only a quadrant is required for finite element analysis. The finite element model is shown in Fig.(3). Four-node isoparametric elements are used for the displacement field and the temperature field, and constant elements are used for the pressure field. Seventy-five elements in total have been used for computation. The displacement d shown in Fig.(3) is used as a control variable. The nonlinear equations are solved using a combination of Newton iteration and line search optimization.

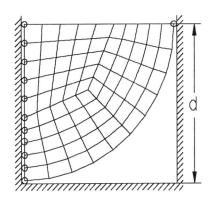

Figure 2 The finite element model the rubber seal

11

The following material coefficients used in the calculations are thought to be representative of an unfilled natural rubber: $C_1=0.293(MPa)$, $C_2=0.177(MPa)$, $\kappa=1410(MPa)$, $c_e=1960(J/kg \cdot K)$, $\alpha=7.5\times10^{-4}(1/K)$, $k=0.14(W/M \cdot K)$, and $\rho_0=916(kg/M^3)$. The remaining parameters are: $k_H=20(GPa)$, $e_r=0.01(cm)$, $\alpha_k=10^7(1/cm)$, $\alpha_h=0.1(W/cm \cdot K)$, and the ambient temperature $T_r=500(K)$.

A special purpose finite element code has been developed based on the foregoing formulations. Computation has been performed for the rubber o-ring seal shown in Fig.(1). The numerical results are shown in Fig.(3)-Fig.(9), in which the units for pressure contours, temperature contours, and geometric dimensions are $10(kPa)$, $1(K)$, and $1(cm)$, respectively. The numerical results indicate that the pressure inside the rubber o-ring seal increases rapidly with increasing temperature, especially after the seal is fully confined. The code is validated numerically in the following procedure: (i) find the pressure distribution and the reaction force from the rigid well to the elastomeric seal for the case (a) in whcih the elastomeric seal is confined; (ii) use the reaction force found in the case (a) as the external force to find the pressure distribution for the case (b) in which the elastomeric seal is not confined. The numerical simulation shows the pressure distribution is the same for the case (a) and the case (b), thereby validating the finite element algorithm.

DISCUSSION AND CONCLUSIONS

A previous formulation for coupled thermomechanical response of near-incompressible elastomers has been extended to the case in which confined elastomeric seals are submitted to thermal and mechanical loads. In the current investigation, traction and heat flux are assumed to be prescribed on the deformed surface and Fourier's law for heat conduction is expressed in terms of the deformed coordinates. Numerical experiments show the continuously nonlinear elastic foundation model used in the current investigation has advantages over the conventional bilinear model which is used widely in commercial finite element codes. The numerical results indicate that the pressure inside the rubber o-ring seal increases rapidly with increasing temperature, especially after the seal is fully confined.

REFERENCES

Lin, Baojiu, 1996, "Finite Element Methods for Thermohyperelastic Bodies under Non-classical Boundary Conditions", Ph. D. dissertation, University of Central Florida, Orlando.

Nicholson, D. W., 1995, "Tangent Modulus Matrix for Finite Element Analysis of Hyperelastic Materials", ACTA MECHANICA, Vol. 112, pp. 187-201.

Nicholson, D. W. and Lin, B., 1996a, "Thermohyperelastic Con -
stitutive Model for Near-Incompressible elastomers", ACTA MECHANICA, Vol. 116, pp. 15-28.

Nicholson, D. W. and Lin, B., 1996b, "Finite Element Method for Thermomechanical Response of Near-incompressible Elastomers", ACTA MECHANICA, in press.

Nicholson, D. W. and Lin, B., 1996c,"Incremental Finite Element Equations for Thermomechanical Response of Elastomers: Effect of Boundary Conditions Including Contact", ACTA MECHANICA, in press .

Nicholson, D. W., Nelson, N. W., Lin, B., and Farinella, A.,1996d, "Finite Element Analysis of Hyperelastic Components", Applied Mechanics Reviews, in review.

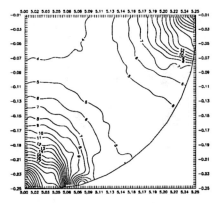

Figure 3 Pressure contours $\Delta d=-0.006(cm)$ and time t=10(sec)

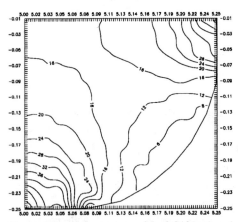

Figure 4 Pressure contours $\Delta d=-0.012(cm)$ and time t=20(sec)

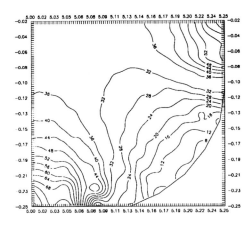

Figure 5 Pressure contours Δd=-0.018(cm) and time t=30(sec)

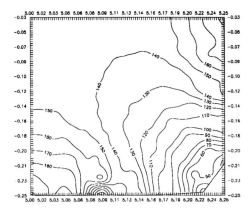

Figure 7 Pressure contours Δd=-0.03(cm) and time t=50(sec)

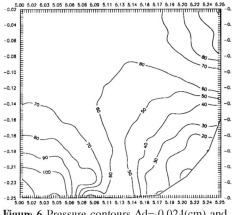

Figure 6 Pressure contours Δd=-0.024(cm) and time t=40(sec)

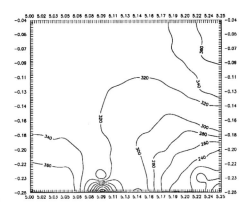

Figure 8 Pressure contours Δd=-0.036(cm) and time t=60(sec)

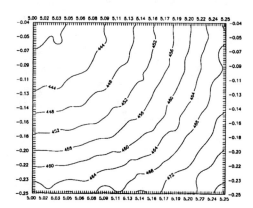

Figure 9 Temperature contours Δd=-0.036(cm)
and time t=60(sec)

PVP-Vol. 356, Integrity of Structures and Fluid Systems,
Piping and Pipe Supports, and Pumps and Valves
ASME 1997

LOCAL PRESSURE STRESSES ON LATERAL PIPE-NOZZLE WITH A
45° DEGREE ANGLE INTERSECTION

James J. Xu
Mechanical Engineering
New Jersey Institute of Technology
Newark, NJ

Benedict C. Sun
Engineering Technology
New Jersey Institute of Technology
Newark, NJ

Bernard Koplik
Mechanical Engineering
New Jersey Institute of Technology
Newark, NJ

ABSTRACT

This paper presents a comprehensive analysis of the local pressure stress data around a pipe-nozzle juncture with a 45 degree angle of intersection. The resulting circumferential and longitudinal stresses on the pipe are first normalized as local stress factors which are then plotted as functions of beta, ß, (nozzle mean radius/pipe mean radius) and gamma, γ, (pipe mean radius/pipe thickness). For the numerical computation of the stresses, the range of beta, ß, is from 0.1, to 1.0, and gamma, γ, from 10 to 300. Through the ALGOR finite element analysis program, comprehensive convergence studies were made for boundary parameters, such as α_p (pipe length/pipe mean radius) and α_n (nozzle length/nozzle mean radius), the optimized numbers of nodes around the pipe-nozzle juncture, and total number of elements of the model.

Twelve (12) plots of local pressure stress factor plots are provided in this paper which allows local stress calculation at all eight locations around the pipe-nozzle juncture that may be combined with stresses resulting from other external loadings as described by the Welding Research Council Bulletin # 107. In this paper a numerical example for the pressure stresses with beta of 0.7 and gamma of 100 is given.

NOMENCLATURE

α_p = pipe length/pipe mean radius
α_n = nozzle length/nozzle mean radius
ß = nozzle mean radius/pipe mean radius

θ = angle between pipe and nozzle axes on the symmetric plan of the pipe-nozzle
L_p = length of pipe
L_n = length of nozzle
p = internal pressure
R_p = pipe mean radius
R_n = nozzle mean radius
t_p = pipe thickness
t_n = nozzle thickness

Subscripts

p = pipe, main shell
n = nozzle, branch

INTRODUCTION

Pipe tees and lateral connections are essential components in process and power generation facilities for functional purposes. The lateral nozzle makes an elliptical opening on the pipe or vessel surface which has a higher stress concentration than the standard 90° nozzle. These high local stresses at the juncture cause major safety concerns especially in nuclear power piping design.

The commonly used laterals have the intersection angles of 30, 45 and 60 degrees, respectively. Considerable literature exists on the subject of local stresses of pipe-nozzle due to internal pressure and other external loadings, but they are limited to the 90 ° intersection case. The fundamental difficulty in the analysis of a lateral pipe-nozzle is that it is not axisymmetric and the curve of intersection is a non-geodesic curve. This paper presents a comprehensive analysis of the data of local stress factors for the 45 degree pipe-nozzle connection as shown in Figure 1. In the figure, points A and B are designated as outside and inside crotch points, respectively.

Using the ALGOR finite element analysis package, the pipe-nozzle juncture is simulated by using a full pipe-nozzle model. To ensure proper convergence of the numerical results on the local stresses, comprehensive studies are made to optimize the models with 96 nodes on the pipe-nozzle junction, and the values for the geometry parameter are 10.0 for α_p (pipe length / pipe mean radius) and 5.0 for α_n (nozzle length / nozzle mean radius). These values ensure that boundary conditions at the end of the pipe and nozzle will not effect the accuracy of the numerical results. Assuming that the membrane pressure stresses in the pipe and the nozzle are identical, the nozzle thickness is proportional to the pipe thickness with a factor beta, ß, i.e. $t_n = ßt_p$.

To provide a comprehensive range of local stress results for design engineers and stress analysts, this paper presents twelve plots of local stress factors for both the circumferential and longitudinal stresses at points A, B, and C, respectively, as shown in Figure 2. In these plots, the geometrical parameter ß (nozzle mean radius / pipe mean radius) ranges from 0.1 to 1.0 with an increment of 0.1 and the γ (pipe mean radius / pipe thickness) ranges from 10 to 300 in ten randomly selected intervals. The local stress factors are defined by normalizing the resulting local stresses by the applied internal pressure value. These stress factor plots are shown in Figures A1 through A12.

THREE-DIMENSIONAL FINITE ELEMENT MODEL

Ten 3-D finite element models are generated by a well developed finite element analysis package, ALGOR, with each specific ß value from 0.1 to 1.0. The pipe-nozzle system here is modelled by using plate/shell elements based on 3-node and 4-nodes. Material properties, such as Young's modulus, Poisson's ratio, thermal expansion coefficient and the material density, are assigned to the elements.

For the analysis, the following assumptions are made:
1. The material is assumed to be homogeneous and isotropic.
2. The resulting stresses are within the proportional limit of the material and obeys Hooke's law.
3. The influence of self-weight is neglected.
4. There are no transitions, fillets, or reinforcing pad at the junction.
5. In the pipe-nozzle model, the boundary conditions in each case do not significantly effect the results of the computation since the parameters α_p (pipe length / pipe mean radius) is assigned as large as 10.0 and α_n (nozzle length / nozzle mean radius) is assigned as large as 5.0.

To simulate the true pipe-nozzle geometry, full finite element models are employed with the symmetric plane (X-Z plane). The number of elements is approximately 4000 to 5000 for the whole finite element models, which is required to develop large number of elements and generate sufficient meshes to provide proper convergence to the stress results.

For the convergence requirement of the finite element method, several models with different numbers of elements, node numbers around the intersection, geometric parameters, and boundary conditions have been studied.

Convergence Studies

The convergence of the finite element models have been carefully studied including the following:

1. The number of the elements for the 3-D finite element models.

The more the number of nodes and elements, the more the accuracy of the results for the finite element model, but more running time will be required. The optimum number of elements for the models in this work is between 4000 - 5500. This study has shown that further increase in the number of elements will only improve the stress results by less than 0.8 percent and the computer running time would be increased substantially.

2. The number of the node points at the juncture of pipe-nozzle.

Ha (1995) has shown that as the number of node points on the pipe-nozzle juncture increases to 96 the stresses will converge asymptotically when the pipe-nozzle intersects with a 90 degree angle. For the 45 degree pipe-nozzle junction, this study also shows that the same 96 node points on the juncture will have the stress results converge asymptotically. This conclusion is based on a typical pipe-nozzle model with ß = 0.5, γ = 50 and the numbers of node points varied from 72 to 112.

3. The optimum values of α_p and α_n.

The boundary parameters, α_p (pipe length/pipe mean radius) and α_n, (nozzle length/nozzle mean radius) should be large enough to obtain the convergence of the solution of various stresses. For the 45 degree pipe-nozzle juncture, studies have shown that $\alpha_p = 10$ and $\alpha_n = 5$ are the optimum values such that the boundary conditions would not have any significant effect on the solution of the stresses at the pipe-nozzle juncture due to internal pressure.

4. Boundary conditions.

In the real pipe-nozzle system, both pipe and nozzle are considered as a closed end system which means the local pressure stresses at the pipe-nozzle juncture are superimposed with the membrane pressure stresses. Meanwhile, in order to prevent the thermal expansion stresses from occurring, the vessel and nozzle are modelled with simply supported ends. It is believed that the above mentioned models should more closely simulate the real piping system.

From the 3-D finite element models used in this study, stresses away from the intersection area approach *PR/T* in the circumferential direction and approach *PR/2T* in the logitudinal direction when the system has closed ends and the boundary conditions are simply supported. This also indicates the real situation that the local stresses are no longer effected by the stress field at the location away from the nozzle area.

Normalization Studies

Normalization studies have verified the validity of using ß, (R_n/R_p) and γ, (R_p/t_p) as the geometric parameters under internal pressure. Several cases of normalization studies have shown that normalizing the pressure stress factor by a randomly selected applied internal pressure is valid with the same geometric parameters, i.e. ß, γ. The local stresses from models of different size but with the same geometric parameters, such as ß, γ, α_p and α_n and which are under the same internal pressure are identical.

TABLE 1: GEOMETRIC PARAMETERS AND DIMENSIONS OF THE EXAMPLE

α_p = Pipe length / Pipe mean radius	10
α_n = nozzle length / nozzle mean radius	5
ß = Nozzle radius / Pipe mean radius	0.7
γ = Pipe radius / Pipe thickness	100
L_p = Pipe length	> 250 in
R_p = Pipe mean radius	25 in.
L_n = Nozzle length	> 87.5 in.
R_n = nozzle mean radius	17.5 in.
t_p = Pipe thickness	0.25 in.
t_n = nozzle thickness	0.175 in.

TABLE 2: LOCAL STRESSES FOR THE EXAMPLE

Data point	Stress factor	Stress, psi	From figure
Longitudinal stress at A_U	867.72	86772	A1
Longitudinal stress at A_L	-640.25	-64025	A2
Circumferential stress at A_U	946.20	94620	A3
Circumferential stress at A_L	499.43	49943	A4
Longitudinal stress at B_U	1706.36	170636	A5
Longitudinal stress at B_L	-1363.34	-136334	A6
Circumferential stress at B_U	2473.91	247391*	A7
Circumferential stress at B_L	1418.72	141872	A8
Longitudinal stress at C_U	-138.20	-13820	A9
Longitudinal stress at C_L	-413.04	-41304	A10
Circumferential stress at C_U	179.62	17962	A11
Circumferential stress at C_L	-209.55	-20955	A12

* *maximum local pressure stress*

ATTENUATION OF LOCAL STRESSES AWAY FROM THE PIPE-NOZZLE JUNCTURE

To study the local stresses effect away from the nozzle, a typical pipe-nozzle model is adopted with α_p = 10, α_n = 5, nozzle radius, R_n = 10 in. , pipe radius, R_p = 20 in. and the pipe thickness is 0.4 in. which yields ß = 0.5, γ = 50. The local pressure stress factors are plotted as a function of x which is the distance from the center of nozzle. Since the nozzle and pipe intersect with 45°, the point A or B is $1.414R_n$ from center of the intersection. From the plots of Figures 1B to 8B, the stress factors from point A or B decrease very fast within a distance of half of $1.414 R_n$. The local stresses approach to the membrane stress values when x approximately reaches twice the value of $1.414R_n$. This agrees with the theory of reinforced openings for the design of reinforcement in the nozzle area as suggested by Harvey (1991).

NUMERICAL EXAMPLE

Example: A pipe with 50.25 in. outside diameter, 0.25 in. thickness, is intersected by a 35.175 in. nozzle with 0.175 in. thickness. The internal pressure is 100 psi. In this example, the mean radius of the pipe, R_p = 25 in., the mean radius of nozzle is R_n = 17.5 in.. Assume that any other nozzle, trunnion, or pipe bend is at least 125 in. away from this nozzle and the nozzle has a minimum length of 87.5 in. All detail geometric information is listed in Table 1, the results are listed in Table 2.

In this example, the circumferential membrane pressure stress is

$$\sigma_c = \frac{pR}{T} = \frac{100 \times 25}{0.25} = 10,000(psi)$$

which is 24.73 times less than the maximum local stress located at the inside crotch point B.

When the elastic modulus is different from 30×10^6 psi, the new local pressure stress may be obtained by multiplying the ratio of new modulus to 30×10^6 psi.

CONCLUSIONS

From the plots of stress factors, Figures A1 to A12, one concludes that:

1. In most cases, the increase of the parameter, γ, (R_p/t_p) makes the local pressure stress higher. It is known from the WRC Bulletin No. 107 (1965, 1984) that when ß increases, the local bending stress decreases and the membrane stress increases. One may conclude that the membrane component is the major contributor to the local pressure stresses when the shell is very thin.

2. The highest local pressure stress occurs at the inside crotch point, B, on the outside surface of the pipe in the circumferential direction. When ß is larger than 0.3, the stresses increase linearly with ß.

3. The circumferential stresses at points A and B are always in tension. However, in the longitudinal direction, these stresses are in tension on the outside surface and in compression on the inside surface.

4. At point C, the local stresses on the inside surface are under compression in both longitudinal and circumferential directions, while on the outside surface of the pipe, the stresses in the circumferential direction are always in tension, but the longitudinal stresses may change from tension to compression when γ and ß increase, as shown in Figure A9.

REFERENCES

Wichman, K. R., A. G. Hopper, and J. L. Mershon. 1965. "Local Stresses in Spherical and Cylindrical Shells Due to External Loading," Welding Research Council Bulletin No. 107.

Mershon, J. L., K. Mokhtarian, G. V. Ranjan, and E. C Rodabaugh. 1984. " Local Stresses in Cylindrical Shells Due to External Loadings on Nozzle-Supplement to WRC No. 107," Welding Research Council Bulletin No. 297.

Ha, J. L., B. C. Sun, and B. Koplik. 1995. " Local Stress Factors of Pipe-Nozzle Under Internal Pressure," *Nuclear Engineering and Design*, 157, pp. 81 - 91.

Harvey, J. F. 1991. *Theory and Design of Pressure Vessel*, Second edition, Van Nostrand Reinhold, N. Y., pp. 409 - 412.

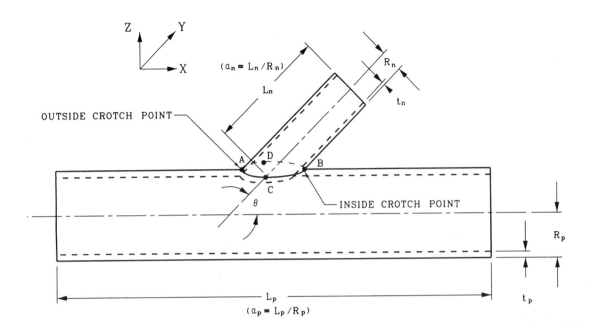

Figure 1: Pipe-nozzle configuration

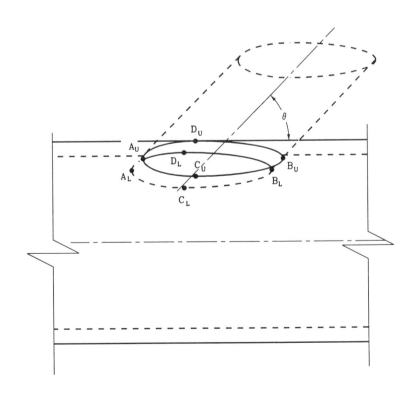

Figure 2: Location of stress data points

19

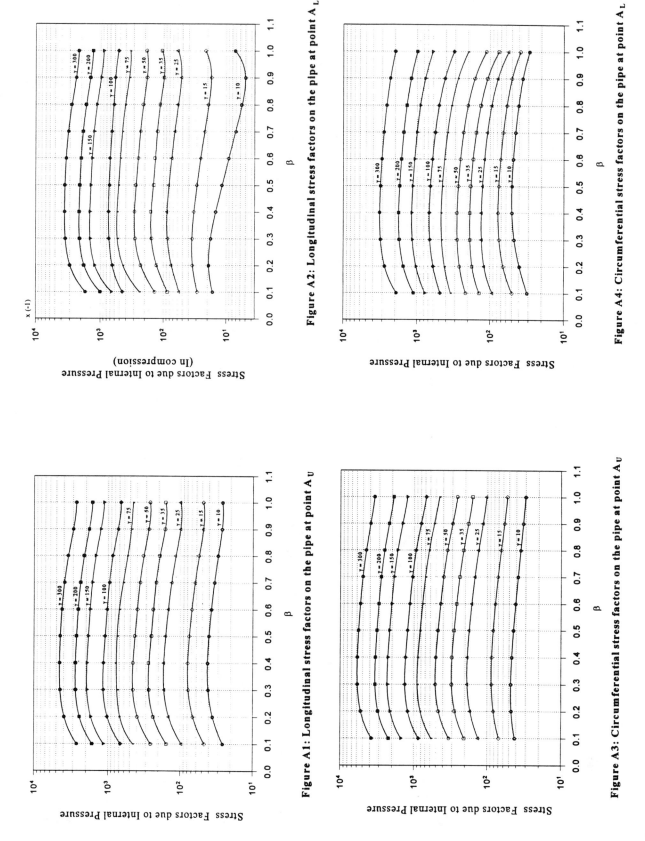

Figure A1: Longitudinal stress factors on the pipe at point A_U

Figure A2: Longitudinal stress factors on the pipe at point A_L

Figure A3: Circumferential stress factors on the pipe at point A_U

Figure A4: Circumferential stress factors on the pipe at point A_L

20

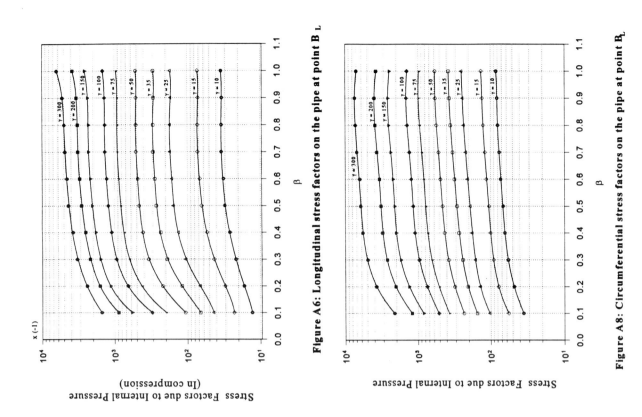

Figure A5: Longitudinal stress factors on the pipe at point B U

Figure A6: Longitudinal stress factors on the pipe at point B L

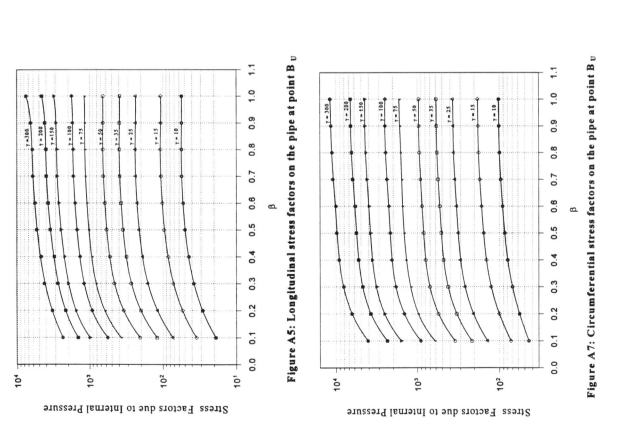

Figure A7: Circumferential stress factors on the pipe at point B U

Figure A8: Circumferential stress factors on the pipe at point B L

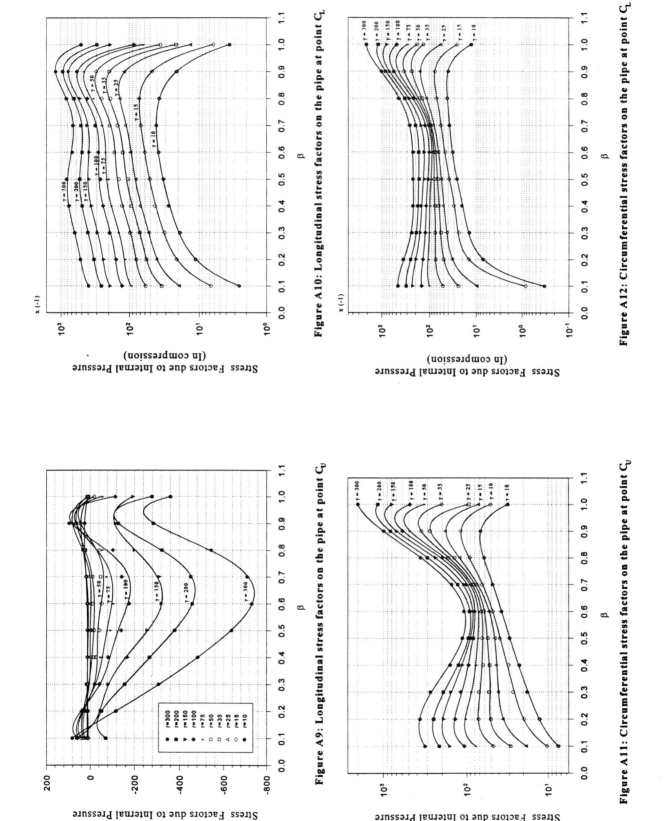

Figure A9: Longitudinal stress factors on the pipe at point C_U

Figure A10: Longitudinal stress factors on the pipe at point C_L

Figure A11: Circumferential stress factors on the pipe at point C_U

Figure A12: Circumferential stress factors on the pipe at point C_L

22

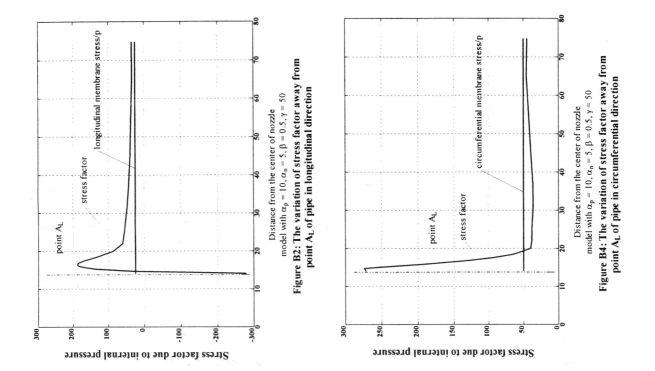

Figure B1: The variation of stress factor away from point A_U of pipe in longitudinal direction

model with $\alpha_p = 10$, $\alpha_n = 5$, $\beta = 0.5$, $\gamma = 50$

Distance from the center of nozzle

Figure B2: The variation of stress factor away from point A_L of pipe in longitudinal direction

model with $\alpha_p = 10$, $\alpha_n = 5$, $\beta = 0.5$, $\gamma = 50$

Distance from the center of nozzle

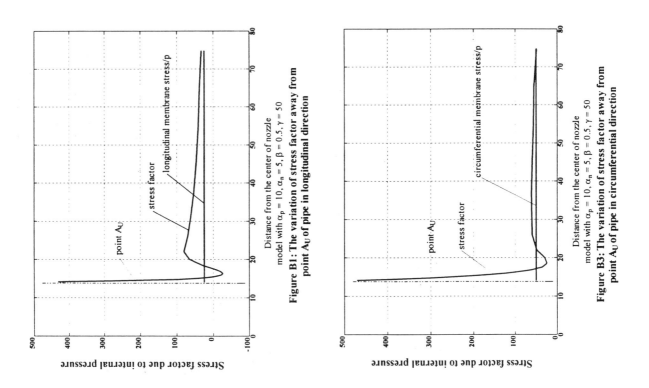

Figure B3: The variation of stress factor away from point A_U of pipe in circumferential direction

model with $\alpha_p = 10$, $\alpha_n = 5$, $\beta = 0.5$, $\gamma = 50$

Distance from the center of nozzle

Figure B4: The variation of stress factor away from point A_L of pipe in circumferential direction

model with $\alpha_p = 10$, $\alpha_n = 5$, $\beta = 0.5$, $\gamma = 50$

Distance from the center of nozzle

23

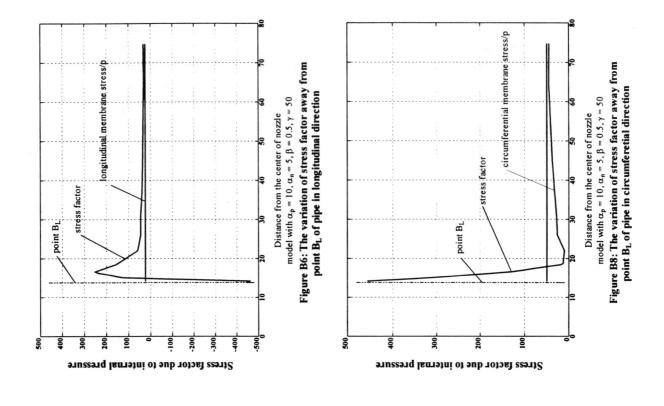

Figure B5: The variation of stress factor away from point B_U of pipe in longitudinal direction
model with $\alpha_p = 10$, $\alpha_n = 5$, $\beta = 0.5$, $\gamma = 50$

Figure B6: The variation of stress factor away from point B_L of pipe in longitudinal direction
model with $\alpha_p = 10$, $\alpha_n = 5$, $\beta = 0.5$, $\gamma = 50$

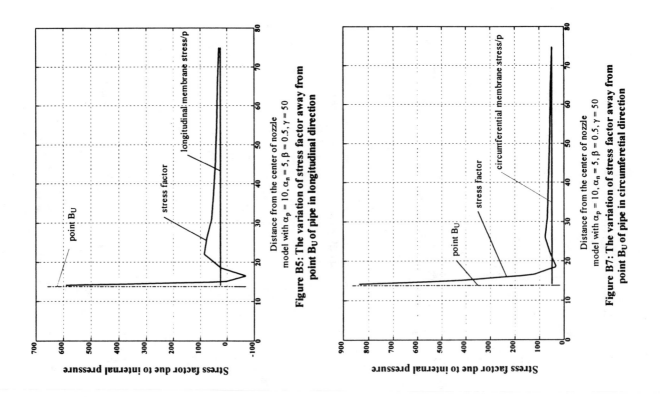

Figure B7: The variation of stress factor away from point B_U of pipe in circumferetial direction
model with $\alpha_p = 10$, $\alpha_n = 5$, $\beta = 0.5$, $\gamma = 50$

Figure B8: The variation of stress factor away from point B_L of pipe in circumferetial direction
model with $\alpha_p = 10$, $\alpha_n = 5$, $\beta = 0.5$, $\gamma = 50$

24

PVP-Vol. 356, Integrity of Structures and Fluid Systems,
Piping and Pipe Supports, and Pumps and Valves
ASME 1997

INVESTIGATION OF INELASTIC STRESS CONCENTRATIONS
AROUND NOTCHES PART I: ANALYTICAL STUDY

Grant B. Stephenson
Gerald H. Lindsey
Young W. Kwon
Naval Postgraduate School
Monterey, California

ABSTRACT

A study of local conditions at notches loaded into the plastic
region is undertaken using both analytical and experimental
means. Part I of the study which focuses upon the analytical work
is reported in this paper wherein nonlinear finite element
solutions are used as the principal analytical thrust. The proposal
of Glinka to determine the local stresses and strains at the notch
root by assuming that in the plastic region the strain energy
density can be approximated with the elastic value is evaluated.
Finite element software and procedures are proved by comparison
with classical plasticity solutions, strain energy densities are
calculated and compared, stress and strain predictions using the
Glinka postulate are compared to those predicted using the
Neuber postulate, which are in turn compared to the finite
element solutions as a standard. It is found that the actual stresses
and strains are bounded by the two models and the average value
between the two comes very close to the actual values calculated.
Errors in stresses are found to be relatively small using the two
models, but errors in strains were found to be significant in the
examples studies.

Part II of the study focuses on the experimental work, which
extends the study, evaluation and comparison of Glinka's and
Neuber's postulates into the realms of metal composites and out
of plane bending.

INTRODUCTION

Critical areas of failure, or fatigue crack initiation, usually
occur at stress concentrations, such as those occurring at notches
or rivet holes. Current fatigue calculations can estimate the cyclic
life of a component given the range of stress and strains it will
undergo; however, for those calculations to be accurate, precise
stress and strain figures are required at the notch. This is
especially true when the cyclic fatigue calculations cover tens of
thousands of cycles, and any error gets multiplied many-fold.
Thus, accurate calculation of the stress at a notch root is the first
step in an accurate fatigue life estimation.

Given a defined geometry and applied loads, K_t, the ratio of
local notch root stress σ to the far-field stress S, remains constant
as long as σ and S remain in the linear range of the stress-strain
curve. Difficulty arises in calculating the stresses for a geometry
such as a notch or a hole when local yielding occurs.   A finite
element
analysis is possible, but this is expensive and time consuming,
especially when used in fatigue calculations which involve tens of
thousands of runs. Stephenson [Ref. 1] has studied this problem
and a more detailed accout of the work presented in this paper is
given by him.

In 1961, H. Neuber [Ref. 2] derived a relationship for
determining stresses and strains at a notch that has been loaded
into the plastic range. Although Neuber's derivation involved a
notch loaded in antiplane shear, it has been widely applied to
general notch problems. By the 1980's the Neuber method had
been adopted by virtually all fatigue analysts; however, over the
years since Neuber's original paper, and ever since it has come
into prominent use, many investigators have proposed alternative
means.

One proposal by Glinka et al. [Ref. 3, 4, 5, 6] was based upon
the concept that the strain energy density of the material in the
yielded zone in the vicinity of the notch is virtually the same as
the strain energy density when the material is elastic. This is
represented below in Figure 1 where W_e is the strain energy
density assuming an elastic material and equals the area under the
linear curve, and W_p is the strain energy density for an elasto-
plastic material and equals the area under the nonlinear curve.
This conjecture results in being able to calculate the stress
concentration factor in the elasto-plastic zone from the strain
energy density of the elastic model. This proposal is based on
Glinka's reasoning that for local plastic yielding, there is a
relatively large volume of material in the elastic region
surrounding the plastic zone. Glinka continued to work on his
proposal through the 1980's, and published several papers
applying his model to plane stress and plane strain problems.

In 1992 W. N. Sharpe, Jr., C. H. Yang, and R. L. Trengoning [Ref. 7] evaluated the Glinka relations with that of Neuber's for various plane strain and plain stress configurations using experimental data at the notch root. Their conclusions were mixed, and they stated that some cases were better predicted by the Glinka model, while others were mere accurately predicted by the Neuber model.

Drawing upon nonlinear finite element techniques this study evaluates the proposal that the strain energy density in the plastic zone is equal to that calculated on the basis of an elastic solution, not only at the notch root, but throughout the plastic zone of the model. Elastic and elasto-plastic problems with closed form solutions, along with previously published experimental data are used to verify the finite element modeling, which is then used to calculate strain energy density. Assessments are made of its impact upon local stresses and strains at the notch root.

NEUBER'S MODEL

Neuber proposed that at the notch root under plastic yielding, the elastic stress concentration factor is the geometric mean of the stress concentration factor and the strain concentration factor, as shown in Equation (1) [Ref. 2]. When the far-field stress and strain, S and e, are in the linear range, this can be rewritten as:

$$\left(K_t S_n\right)^2 = E\sigma\varepsilon \qquad (1)$$

There are two unknowns here, σ and ε, the local stress and strain. To solve for them, the Ramberg-Osgood expression for uniaxial stress was used and the Neuber relation became

$$\frac{(K_t S_n)^2}{E} = \frac{\sigma^2}{E} + \sigma\left(\frac{\sigma}{K}\right)^{\frac{1}{n}} \qquad (2)$$

While Neuber's rule has been well established as an engineering tool to calculate notch stresses and strains, it has been shown to overestimate these values [Ref. 3].

GLINKA MODEL

Glinka proposed that the energy density at the notch root, calculated on the basis of elasto-plastic constitutive laws, is equal to that based on linear elastic constitutive laws for equivalent far-field loading. Elastic strain energy density at the notch and the far-field are calculated using a linear elastic stress-strain relationship.

$$W_\sigma = \int_0^e E\varepsilon d\varepsilon = \frac{E\varepsilon^2}{2} = \frac{\sigma^2}{2E} \qquad (3)$$

Relating these values to the theoretical stress concentration factor gives

$$K_t^2 = W_\sigma / W_s \qquad (4)$$

However, Glinka's hypothesis is that the strain energy density at the root will result in the same value, regardless if calculated for a linear elastic or an elasto-plastic material. Therefore, this ratio remains constant, even when local yielding occurs at the notch root.

For a Ramberg-Osgood nonlinear stress-strain relationship, W_σ is found to be

$$W_\sigma = \frac{\sigma^2}{2E} + \left(\frac{\sigma}{1+n}\right)\left(\frac{\sigma}{K}\right)^{\frac{1}{n}} \qquad (5)$$

Substituting Equations (3) and (5) into Equation (4) results in:

$$\frac{\left(K_t S_n\right)^2}{E} = \frac{\sigma^2}{E} + \frac{2\sigma}{1+n} + \left(\frac{\sigma}{K}\right)^{\frac{1}{n}} \qquad (6)$$

By inspecting Equation (6) from Glinka and Equation (2) from Neuber, one can observe that the only difference is the factor of $2/(1+n)$ in the strain energy density model. Since n is less than 1, this term is greater than 1, and for the left side of these two equations to be equal, the local stress in Equation (6) must be less than the local stress in Equation (2). Likewise, if the local stresses in the strain energy model are lower than those in the Neuber model, so will be the local strains. In fact, Glinka states that the Neuber model has been shown to overestimate the local stresses and strains [Ref. 3].

VERIFICATION OF FINITE ELEMENT METHOD

To evaluate the strain energy density concept for calculating stress concentrations in the plastic zone, a finite element model was used to obtain the stress and strain data as an input to the calculations. However, verification of the finite element, software and procedures was essential to ensure that a legitimate analysis was performed. The finite element program used was I-DEAS Master Series™ Release 1.3 and 2.1 Quadrilateral plane stress and plain strain elements were employed, and a non-linear stress-strain analysis was performed.

The general approach was to create a three dimensional model to represent the physical specimen. Next, an element mesh was created by specifying the type of element to be used and the number of elements on each side of the surface being meshed. All plane stress and plane strain analyses used an eight node quadrilateral element. These elements were two-dimensional, with nodal degrees of freedom consisting of translation in the x and y directions, and rotation about the z axis. Only one face of the model was required to be meshed. A mesh refinement routine was used to refine the element shapes to reduce skewing and stretching.

For each problem, a mesh study was performed to determine the optimal mesh size and distribution to use. This involved running several finite element models with varying mesh sizes and distributions, and determining at what point further mesh refinements did not effect the solution.

Two additional factors effect the non-linear portion of the analysis: the non-linear stress-strain relationship and the iteration procedure. For a non-linear stress-strain analysis, the I-DEAS Master Series™ program uses input data points to model the stress-strain curve. The program assumes constant stiffness between each point, breaking the curve up into individual linear sections. A maximum of twenty points may be entered. To ensure a smooth curve, all twenty points were used to define the stress-strain curve, with denser groupings employed at locations of highest curvature.

Two basic types of comparisons were made: (1) finite element to analytic solutions, and (2) finite element to experimental data. The first type provides a comparison to an exact solution using the same assumptions that are employed in the finite element theory, and the second will indicate how well the finite element method models actual mechanical behavior. The critical area in this case is the elasto-plastic relationship, or how well the Prandtl-Reuss equations represent this relationship. This includes using effective stresses effective strains and a uniaxial stress-strain behavior to model three dimensional constitutive and yield behavior.

Full stress field measurements in solid mechanics must rely on surface techniques and assume plane stress or plane strain conditions. There are also approximations necessary in obtaining the elasto-plastic measurements, such as reducing the data via the Prandtl-Reuss equations. Another disadvantage of using experimental data is the percent error introduced when the measurements are made; however, an attempt was made to compare the finite element results to previously published experimental results.

Verification of Elastic Finite Element Model

The first step in the finite element modeling was to ensure that the software could obtain accurate results for a notched type of geometry under linear elastic loading. An elliptical hole in an infinite thin plate was used for this comparison. Brown [Ref. 8.] gives the complete stress field for unaxial loads.

To approximate an infinite plate, a plate 40 inches by 40 inches containing a central elliptical hole with a semi major axis of a 1 inch and a semi minor axis of 0.5 inches was used. A farfield stress of S = 1000 psi was applied. Due to symmetry, one quarter of the block was modeled (20 inches by 20 inches).

Comparison of the finite element results with that of the analytical solution shows very good agreement between the two. Figure 2 shows the tangential stress along the boundary of the ellipse plotted against the x-axis location. The finite element results coincide with the analytic solution across the complete boundary. The normalized error was generally less than 1.0%, even at the point of maximum stress ($x = 1$). The other stresses compared equally favorably with the analytic solution.

Verification of Elastoplastic Finite Element Model

The first elastoplastic comparison of the I-DEAS™ Software was on a long narrow block to verify the input of the stress-strain curve. A finite element mesh of 3 elements by 100 elements with the outer dimensions 1.0 m by 0.03 m, was used. This simple case illustrated that the elasto-plastic solution converges to the input stress-strain curve. Figure 3 shows the stress-strain curve used, along with the FEM calculations.

Infinite Plate with a Circular Hole

Figure 4 shows the problem selected for this verification. To obtain the elasto-plastic stresses and strains, both incremental and deformation theories have been applied to this problem. [Ref. 9, 10, 11, and 12]. Tuba [Ref. 10] and Chakrabarty [Ref. 12] show that both these solutions are identical. The solution below is taken from Chakrabarty.

Applying the governing equations in polar form for axisymmetric loading, one obtains the stress components in terms of undetermined constants. By enforcing continuity of stress at the elastic-plastic boundary and applying the loading stress, σ, at the infinite boundary, the elastic stress distribution is found to be

$$\sigma_r = \sigma - \frac{\sigma_0 c^2}{\sqrt{3}r^2}\sqrt{1 - \frac{\sigma^2}{\sigma_0^2}} \qquad \sigma_\theta = \sigma + \frac{\sigma_0 c^2}{\sqrt{3}r^2}\sqrt{1 - \frac{\sigma^2}{\sigma_0^2}} \qquad (7)$$

where σ_c is the uniaxial yield stress and c is the radial location of the elastic-plastic interface. It is not possible to express the stress components in the plastic region in such simple terms, but a very clever method of solution leads to

$$\sigma_r = \frac{2}{\sqrt{3}}\sigma_e \sin\phi \qquad \sigma_\theta = \frac{2}{\sqrt{3}}\sigma_e \cos\left(\frac{\pi}{6} - \phi\right) \qquad (8)$$

From equilibrium and compatibility conditions, σ_e is found to be

$$\sigma_e = K\sigma\left(\cos\phi + \sqrt{3}\,n\sin\phi\right)^{-n(1+3n)/(1+3n^2)} e^{\left(-\sqrt{3}\frac{n(1-n)}{1+3n^2}\phi\right)} \qquad (9)$$

and $\phi(r)$ is given implicitly by

$$\frac{a^2}{r^2} = \frac{1}{\sqrt{3}}\left(\sqrt{3}\cos\phi - \sin\phi\right)$$

$$x\left(\cos\phi + \sqrt{3}\,n\cos\phi\right)^{\frac{-4n}{1+3n^2}} e^{\left(-\sqrt{3}\frac{1-n^2}{1+3n^2}\phi\right)} \qquad (10)$$

where ϕ is a function of r, and K is a function of the applied stress, the value of ϕ at the elastic-plastic interface, ϕ_c, and the shape of the stress-strain curve, which is defined by the yield stress, σ_0 and the strain hardening coefficient, n.

$$(11)$$

$$K = \frac{\sigma_0}{\sigma}\left\{(1+3n)\frac{\sigma}{2\sigma_0} + \frac{\sqrt{3}(1-n)}{2}\sqrt{1 - \frac{\sigma^2}{\sigma_0^2}}\right\}^{\frac{n(1+3n)}{1+3n^2}} e^{\sqrt{3}\frac{(1-n)}{1+3n^2}\phi_c}$$

$$\phi_c = \sin^{-1}\left(\frac{\sqrt{3}\sigma}{2\sigma_0} - \frac{1}{2}\sqrt{1 - \frac{\sigma^2}{\sigma_0^2}}\right) \qquad (12)$$

The form of the stress-strain relation used was a modified Ludwick curve, which is defined as follows:

$$\sigma = \begin{cases} E\varepsilon & \varepsilon < \dfrac{\sigma_0}{E} \\[2ex] \sigma_0\left(\dfrac{E\varepsilon}{\sigma_0}\right)^n & \varepsilon \geq \dfrac{\sigma_0}{E} \end{cases} \qquad (13)$$

A plot of a family of curves of this form is shown below in Figure 5, where $\sigma_0 = 30,000$, $E = 30.0 \times 10^6$, and n varies from 0 to 0.5.

For a given applied load of $\sigma_r = \sigma$, the plastic stresses in Equation (8) can be obtained by first solving for ϕ_c and K_t as given in Equations (11) and (12). The auxiliary angle ϕ in Equation (10) is solved for a given radius by an iterative method. Finally σ_e is solved for that value of r from Equation (9), which is then substituted into Equation (8) to solve for the component stresses. For a work-hardening material ($n \neq 0$), the strains are then found by Equation (14).

$$\varepsilon_r = \frac{1}{E}(\sigma_r - \nu\sigma_\theta) + \frac{3}{2}\left(\frac{1}{E_s^t} - \frac{1}{E}\right)\left[\sigma_r - \frac{1}{3}(\sigma_r + \sigma_\theta)\right] \qquad (14a)$$

$$\varepsilon_\theta = \frac{1}{E}(\sigma_\theta - \nu\sigma_r) + \frac{3}{2}\left(\frac{1}{E_s^t} - \frac{1}{E}\right)\left[\sigma_\theta - \frac{1}{3}(\sigma_r + \sigma_\theta)\right] \qquad (14b)$$

The plastic secant modulus has been rewritten as follows:

$$\frac{1}{E_s} = \frac{\varepsilon_e^p}{\sigma_e} = \frac{\varepsilon_e^t - \varepsilon_e^e}{\sigma_e} = \frac{1}{E_s^t} - \frac{1}{E} \qquad (15)$$

were E_s^t is the secant modulus of the stress total strain curve.

Two cases were solved analytically and used as comparisons to the finite element solution with equal success. The first involved an elastic-perfectly plastic material that matched that used by Davis [Ref. 9] for his numerical results using incremental theory. The second case used the same elastic properties but changed the plastic property to a work hardening material. Table 1 below shows the material properties and calculated parameters.

The mathematical software, Maple V®, was used to solve for ϕ for each input radius. Once ϕ was found for a specific radius, the calculations of σ_e, followed by σ_r and σ_θ, were straight forward. Maple V® was programmed to make all these calculations at a radial distance corresponding to each node location along the x-axis.

Finite Element Solution

The infinite plate with a circular hole was modeled using a 60 degree section on an annulus with an inner hole radius of $a = 1.0$ and an outer radius of 30. Due to the axisymmetric loading, a variety of finite elements were available that would efficiently model the axisymmetric problem; however, the above geometry with quadrilateral plane stress elements was used in order to verify these elements and the I-DEAS™ software for more general notched geometries.

Results of the FEM were taken along the one line of elements at $\theta = 0$. The radius of the elastic/plastic boundary, c, corresponding to an effective stress of 30,000 psi, was interpolated between nodal points to be 1.5185 inches for case one, and 1.5107 for case two, which compare favorably with the analytic values listed in Table 1. This is an error of only 0.2% for the former case, and 0.59% for the later. Figure (6) shows the finite element stresses with the analytic curves superimposed for the stress distributions for case 2. Very good agreement was obtained with the analytic solution in both cases, and the normalized error was less than 1% for both cases. For the elastic/perfectly-plastic model, the strains in the plastic region

cannot be obtained analytically. However, for case 2, one simply solves Equation (14). The strain distribution for case 2 is shown in Figure (7). As in the case of the stress, the finite element method produced very accurate results.

Comparison to Experimental Data
Finite Element Modeling of Experimental Specimen

Although the availability of recently published full-field stress data for basic notched specimens under tensile loads were difficult to find, two notable works from Durelli and Sciammarella [Ref. 13] and Theocaris and Marketos [Ref. 14] were found. A comparison to the Theocaris and Marketos experiment was chosen. This peak σ_y stress progression occurs due to a multi-axial stress state and given stress distributions allowing for a higher σ_y stress before yielding occurs. This was also the result of all the finite element analyses completed. The test specimens used by Theocaris and Marketos were two sheets of aluminum alloy 57S, one with a hole diameter to width ratio of 1/2, the other 2/3. The later ratio of was chosen.

The stress-strain curve of the Aluminum alloy 57S as given by Theocaris and Marketos was fit with the Ramberg-Osgood equation. Six levels of applied loads were used. The first load resulted in the initial onset of plastic deformation, while the last load creates a plastic zone that extends a distance 3/4 of the hole radius from the edge of the hole and covering approximately 1/3 of the minimum cross section.

Finite Element Results

A comparison of the σ_y stress was made at the minimum section point for all six load sets, and these are shown in Figures (8) and (9). Several differences are noticeable between the FEM results and the Theocaris and Marketos data. First, although Theocaris and Marketos show a peak σ_y stress that progresses inward as the plastic deformation increases, they also show that it initially decreases before reaching a maximum for the last three applied loads. The FEM shows that the maximum value of σ_y moves away from the hole edge, but it also shows that σ_y constantly increases until it peaks. The second difference between the FEM and the Theocaris and Marketos data is the magnitude of the decrease in σ_y near the edge opposite the hole. This results in significant disagreement between the two values, approaching an 80% difference for the first applied load. One test of the accuracy of both results is to determine if equilibrium has been satisfied at the minimum cross section. The stress distribution curves were numerically integrated to determine the resulting force. The values are only for half of the plate, hence the total applied force will be twice these values. Even though these calculations are only approximate, it is easily seen that the FEM has satisfied equilibrium, while the Theocaris and Marketos results have underestimated the stress distribution for the first three load sets. It should also be noted that the photoelastic analysis shows fringes in the regions of strain gradients, and since this is a region of relatively uniform stress and strain, the accuracy of the method degrades.

The finite element analysis generally compared favorably with the experimental data of Theocaris and Marketos. Within 20 mm of the hole edge, the error was less than 10% for all but one load set. Additionally, the results at the edge itself were within 4%. In addition to the experimental errors referred to, the data used in these shortcomings, the FEM analysis provided quantitative results as reasonably as could be expected, and matched the qualitative trends quite well.

Three Dimensional Strain Energy

To investigate the energy density comparison throughout the yield zone, a three dimensional expression for strain energy density is needed.

$$W_o = \int_0^{\varepsilon_{ij}} \sigma_{ij} d\varepsilon_{ij} \tag{16}$$

For the elastic case, this integrates to the familiar

$$W_o^e = \frac{1}{2}\sigma_{ij}\varepsilon_{ij} \tag{17}$$

For the elastic-plastic case, the strain energy density relationship is separated into an elastic term and plastic term:

$$W_o = W_o^e + W_o^p = \int \sigma_{ij} d\varepsilon_{ij}^e + \int \sigma_{ij} d\varepsilon_{ij}^p \tag{18}$$

The first term is equation (16), and the second term can be manipulated to give:

$$W_o = \frac{1}{2}\sigma_{ij}\varepsilon_{ij}^e + \sigma_{ij}\varepsilon_{ij}^p - \int \varepsilon_{ij}^p d\sigma_{ij} \tag{19}$$

The stress-strain relationship for elastic-plastic behavior for deformation theory based on the Prandtl-Reuss equations is given as (20).

$$\varepsilon_{ij} = \left[\frac{1}{2G}\sigma_{ij} - \frac{\nu}{E}\sigma_{kk}\delta_{ij}\right] + \frac{3}{2}\frac{1}{E_s}\left[\sigma_{ij} - \frac{\sigma_{kk}}{3}\sigma_{ij}\right] \tag{20}$$

Where E_s is the secant modulus of the effective stress versus plastic strain curve:

$$E_s = \frac{\sigma_e}{\varepsilon_e^p} \tag{21}$$

and the effective stress, σ_e, and effective plastic strain strain, ε_e^p, are defined as:

Effective Stress: $\sigma_e = \sqrt{\frac{3}{2}s_{ij}s_{ij}}$ (22)

Effective Plastic Strain: $\varepsilon_e^p = \sqrt{\frac{2}{3}\varepsilon_{ij}^p\varepsilon_{ij}^p}$ (23)

Substituting the plastic strain component of the relationship given in Equation (20) into Equation (19), the strain energy density becomes:

$$W_o = \frac{1}{2}\sigma_{ij}\varepsilon_{ij}^e + \sigma_{ij}\left[\frac{3}{2E_s}\left(\sigma_{ij} - \frac{\sigma_{kk}}{3}\sigma_{ij}\right)\right] - \int \frac{3}{2E_s}\left(\sigma_{ij} - \frac{\sigma_{kk}}{3}\delta_{ij}\right)d\sigma_{ij} \tag{24}$$

If the Ramberg-Osgood uniaxial stress-strain curve is rewritten in terms of effective stress and effective strain, and substituting the plastic portion of the effective strain of the Ramberg-Osgood relationship into the definition of the secant plastic modulus, Equation (21), can be rewritten in terms of the effective stress and the Ramberg-Osgood material constants.

$$\frac{1}{E_s} = \frac{\left(\frac{\sigma_e}{K}\right)^{\frac{1}{n}}}{\sigma_e} = \left(\frac{1}{K}\right)^{\frac{1}{n}}\sigma_e^{\frac{1-n}{n}} \tag{25}$$

Substituting this back into equation (24) results in the strain energy density relationship in terms of stresses only (based on the Ramsberg-Osgood stress-strain curve):

$$W_o = \frac{1}{2}\sigma_{ij}\varepsilon_{ij}^e + \sigma_{ij}\left[\frac{3}{2K^{\frac{1}{n}}}(\sigma_e)^{\left(\frac{1-n}{n}\right)}\left(\sigma_{ij} - \frac{\sigma_{kk}}{3}\delta_{ij}\right)\right] \tag{26}$$

$$-\int \frac{3}{2K^{\frac{1}{n}}}(\sigma_e)^{\left(\frac{1-n}{n}\right)}\left(\sigma_{ij} - \frac{\sigma_{kk}}{3}\delta_{ij}\right)d\sigma_{ij}$$

Finite element solutions applied to notched geometries were numerically integrated to calculate the actual strain energy distributions density from this equation. Because of the monotonic nature of the loading in the problems studied, the incremental and deformation theories coalesce and this form of the energy equation applies.

STRAIN ENERGY DENSITY CALCULATIONS

In order to test the validity of the Glinka strain energy density proposal, the strain energy density for a given loading was calculated based on an elasto-plastic material (W_p) and an elastic material (W_e), and comparisons between the two calculations were made. The strain energy density calculations were performed at each node throughout the FEM model, with the plastic strain energy density (W_p) being numerically integrated at each load step by use of a third order integration scheme.

Notch Geometry and Material Selection

Two separate geometries were evaluated, one had a notch radius of 1.0 inch, and a plate width of 6.0 inches (r/D = 16), and the second had a notch radius of 1.0 inch and a plate width of 10.0 inches (r/D = 1/10). These two geometries are shown below in Figure 10, with the shaded regions representing the finite element geometry. For each geometry, plane stress (thin plate) and plane strain (thick plate) analyses were performed. The strain-hardening material used modeled 7075-T6 Aluminum. The yield stress was determined to be 66 ksi, based on a plastic strain offset of 0.002; however, it should be noted that the stress-strain curve departs from linearity at 40ksi. The 20 data points were obtained by curve fitting the Ramberg-Osgood equation to actual stress-strain data from the Military Handbook V [Ref. 15]; then calculating these points from the resulting Ramberg-Osgood equation.

29

Finite Element Modeling

Due to the two lines of symmetry for this geometry, the finite element model was reduced to one quarter of the physical model by applying appropriate constraints along each boundary, as shown in Figure 10. For each configuration, 21 increments were used to increase the loading from an initial nominal stress of 12 ksi to a final value of 49.5 ksi, or 75 percent of the yield stress. This load increment ensured convergence of the finite element solution and provided a small enough step to numerically integrate the strain energy density with reasonable accuracy. The first two load points were in the elastic range, and plastic deformation began at the third load step. The stress concentration factor K_t was determined from the finite element analysis at the first load step, and found to be 2.042 and 2.421 for the narrow and the wide plate, respectively.

Results of Finite Element Analysis

The elastic strain energy density (W_e) at each node was initially calculated at the first load step. To determine W_e at later load steps, this value was simply multiplied by a factor of $(P_i/P_0)^2$, where P_i is the load at the ith step, and P_0 is the initial load. Plastic strain energy density (W_p) was calculated incrementally.

Plane Stress Condition

For both plate widths, the plastic strain energy density was found to be greater than the elastic strain energy density in the vicinity of the notch root. Figure (11) illustrates behavior typical of both plots of W_p at load step 21 under plane stress conditions. It shows that the plastic strain energy density throughout the model has its maximum value at the notch root, but rapidly approaches the far-field value away from the notch. Figures (12) and (13) show W_p for several load steps along the minimum cross section of the plates, with W_e included at the final load step. This shows that not only does W_e give under estimated values at the notch root, but also follows a different distribution shape than W_p. At the notch root itself for the final load step ($S = 0.75\sigma_0$), the difference between W_p and W_e reaches a value of 16.8% and 23.2% for the narrow plate and wide plate respectively. Figure 14 shows a contour values indicate that the actual W_p is higher than that estimated by the Glinka Method. These values are taken at the final load step ($S = 0.75\sigma_0$). In the plastic region, the greatest difference between the two calculations occurs not at the notch root itself, as may be expected, but slightly offset along the x-axis. This maximum error is also on the order of twice the value of that at the notch root. The location of the maximum difference between W_p and W_e corresponds fairly well with the location of the maximum σ_y value. It should be noted that although the contour plots show the maximum global error occurring slightly above the notch near the plate edge, this is also a region of very low to zero strain energy density, and therefore these errors are actually insignificant. These results showed that, as previously stated, the maximum error occurs at the notch root for small plastic yielding, then gradually progresses inward along the x-axis. Global errors for the previous load steps followed the same trends as that of the final load step. Starting with zero error at the initiation of plastic deformation, the regions of significant error (greater than 1%) started at the notch root, and moved inward, while at the same time proceeding at an angle toward the vertical centerline of the plate. This trend also corresponds to the region

of high σ_e and the growth of the plastic zone (this growth is well documented by Theocaris and Marketos [Ref. 16]).

Plane Strain Condition

Results of the strain energy density calculations for the plane strain condition showed better agreement between W_p and W_e than for the plane stress condition. As in the plane stress case, there appears a high peak value at the notch root, with a rapid drop-off to the far-field value. Figures 15 and 16 show W_p for several load steps along the minimum cross section of the plates, with W_e included at the final load step. Closer agreement between the two strain energy densities is shown than for the plane stress condition. The plane strain problem results in a configuration that is physically more constrained than that of the plane stress problem, and hence the amount of plastic growth at the notch under plane strain conditions will be less than that of the plane stress condition. Therefore, on the basis that the strain energy density distribution in the plastic region remains relatively unchanged due to a high volume of elastic material surrounding the plastic region should be even more valid for the plane strain condition than that of the plane stress condition. This was in fact shown to be the case when comparing the finite element plastic strain energy density W_p with that of W_e. As in the case of the plane stress condition, W_p for the plane strain condition is shown to be greater than that of W_e; however, both the amount of error and region of significant error (greater than 1%), is much improved over the plane stress condition. At the notch root itself, the difference between the two calculations was only 7.76% for the narrow plate and 10.6% for the wide plate.

STRESS AND STRAIN CALCULATIONS

Plane Stress Condition

The stress distributions for plane stress for both geometries are shown in Figure 17. These plots depict either σ_y or σ_x along the minimum cross section ($y = 0$) as a function of the nominal loading. As stated earlier, the maximum σ_y value progress inward from the notch root as yielding increases. This can be attributed to the σ_x component, which starts from zero at the notch root and rapidly approaches its maximum value inward from the notch root. This increase in the σ_x component results in a higher allowable σ_y than at the notch root before yielding occurs.

Notch Root Stress and Strain

In addition to comparing the plastic strain energy density W_p to the predicted strain energy density W_e, comparisons were also made between the finite element stress and strain results with those predicted by the Glinka and Neuber methods. If the nominal stresses are high enough then $S \neq Ee$, and Equations (6) and (2) are not valid. They may be modified, however, by using the Ramberg-Osgood equation instead of Hooke's law to determine the nominal strains. For example, the Neuber's Method would become:

$$K_t^2 S_n \left[\frac{S_n}{E} + \left(\frac{S_n}{K} \right)^{1/n} \right] = \frac{\sigma^2}{E} + \sigma \left(\frac{\sigma}{K} \right)^{1/n} \qquad (27)$$

Likewise, the Glinka method would result in:

$$K_t^2 S_n \left[\frac{S_n}{E} + \frac{2}{n+1} \left(\frac{S_n}{K} \right)^{1/n} \right] = \frac{\sigma^2}{E} + \frac{2\sigma}{n+1} \left(\frac{\sigma}{K} \right)^{1/n} \qquad (28)$$

However, when these modified forms were applied to the loading levels analyzed in this study the amount of improvement was a fraction of 1%.

The strain energy density calculations showed that the predicted strain energy density based on elastic material properties (W_e) was less than the actual strain energy density (W_p). From this comparison of strain energy densities, it was known that the Glinka method would under predict the stresses and strains at the notch root. The stress and strains were also predicted based on the Neuber method. As was stated previously the Neuber method has been shown to overestimate the stresses and strains. This was also true for all of the configurations analyzed in this work. Results of the Glinka method under predicting the stresses and strains, and the Neuber method over predicting the stresses and strains are shown in Figure 18.

These plots show the stress and strain concentration factor as a function of the nominal load for the FEM analysis, the Glinka method and the Neuber method. Additionally, a stress based on the average of the Glinka and Neuber determined stresses, and a strain determined from the Ramberg-Osgood equation and this average stress were determined for all configurations. It should be noted that when strain values were based on an average of the Glinka and Neuber methods, they are based on the average of the calculated stresses, and not the average of the strains. The typical errors for the stress and strain predictions for both methods are shown in Figure 19. Note that the strain is more sensitive than the stress for both calculations. This is evident from the fact that for the uniaxial stress-strain curve beyond the yield point, strain is highly sensitive to changes in stress. An example of this sensitivity is plotted in Figure 20 which shows the percent error of the Glinka method predictions with respect to the percent difference in strain energy density. Note that this sensitivity is also dependent on loading condition as it relates to the current stress-strain relationship. For example, as the loading increases, the error in stress prediction appears to asymptotically approach a single value after an initial increase, while the error in strain is almost linear with respect to the error in strain energy density. As a material approaches perfectly plastic, any change or error in W_p will result in a linear change in ε, since σ will approach a constant value.

For the notch geometry and plane stress condition, the Glinka and Neuber method give an upper and lower bound to both the stress and strain predictions. Results based on the average of the stresses of the Glinka and Neuber method are in good agreement with the FEM results. It should be noted after about a S/σ_0 ratio of 0.5 to 0.6, the rate at which the error increases for the Glinka method appears to be constant, while even though the error for the Neuber method continually increases, that rate at which it increases diminishes. This results in the error for the average of these two methods to reach a maximum between a S/σ_0 ratio of 0.54 to 0.64, then decrease as the loading increases. However, it does not appear to asymptote toward zero, but to merely change sign as the error from the Glinka method becomes greater than that from the Neuber method.

Plane Strain Condition

The stress distribution for plane strain results in higher axial stresses when compared to the plane stress condition. This is due to the σ_z stress, which produces a higher hydrostatic pressure for a given loading; thus reducing the amount of plastic growth. The reduced plastic growth likewise results in the stresses increasing at a higher rate than in the plane stress condition. Additionally, the plane strain condition results in higher stress gradients in the vicinity of the notch. As in the plane stress condition, the σ_y stress peaks at a point inward from the notch root as plastic growth occurs. For the plane strain condition, this also occurs for the σ_z stress.

Notch Root Stress and strain Calculations

For plane stress analysis at the notch root, W_p is a function of σ_y and ε_y. For the plane strain condition, however, ε_y is a function of both σ_y and σ_z, and the strain energy density equation does not reduce to the simple uniaxial version. To solve the plane strain problem, Glinka [Ref. 4, 5, 6] uses the transformation as suggested by Dowling, et al. [Ref. 17] that relates the uniaxial stress-strain curve to a plane strain stress-strain curve. From this transformed stress-strain curve, ε_y can be found directly from σ_y. To obtain ε_y, ε_y and σ_y are related to the uniaxial stress-strain curve as shown below:

$$\sigma_y = \frac{\sigma}{\sqrt{1-\mu+\mu^2}} \qquad \varepsilon_y = \frac{\varepsilon(1-\mu^2)}{\sqrt{1-\mu+\mu^2}} \qquad (29)$$

where

$$\mu = \frac{v + E\dfrac{\varepsilon_p}{2\sigma}}{1 + E\dfrac{\varepsilon_p}{\sigma}}$$

The Ramberg-Osgood relationship can then be fitted to these transformed stresses and strains, and new parameters determined. Once these values are found, the stresses at the notch root can then be found by substituting all the normal coefficients with the transformed coefficients in Equation (6) and Equation (2), then use the revised Ramberg-Osgood equation to determine the strains.

Since the results of the strain energy density comparison were much improved for the plain strain condition, it was expected that the stress calculations using the Glinka method would also be improved. This was indeed the case, as is shown in Figure 21, which compares the stress concentration factor as a function of the nominal loading. Note that again an average value of the two methods was calculated, and plotted for comparison. Typical errors in both the stress and strain calculations are shown in Figures 22. These show that the Glinka method provides results three times more accurate than the Neuber method, but that the errors in both methods are quite small. Additionally, strains calculated based on the average stress of the two methods give slightly more accurate values than the Glinka method itself. This process also results in strains that are slightly higher than those of the FEM analysis, vice those of the Glinka method, which are slightly lower than the FEM analysis.

31

CONCLUSIONS

This study has examined the proposal by Glinka that the strain energy density at the notch root is approximately the same regardless if the material is elastic or elasto-plastic. A detailed comparison of the two strain energy densities was performed not only at the notch root, but throughout the field of symmetrical, semi-circular double notched plates. These comparisons were made for both plane stress and plain strain conditions. Strain energy density was calculated based on finite element analyses that had been rigorously tested with analytic solutions and experimental data. The strain energy density was numerically integrated, applying 21 load steps to reach a nominal stress equal to three fourths of the yield stress for each configuration.

In addition to strain energy density calculations, stress and strain calculations based on Glinka's strain energy density proposal and the Neuber method were performed, and compared with the finite element method with the following results:

- The strain energy density in the vicinity of the notch root based on elasto-plastic material properties is higher than the strain energy density assuming elastic only properties.
- The plane strain condition results in better agreement between the two strain energy densities than the plane stress condition.
- At the higher loads, the greatest deviation between the two energies occurs slightly inward from the notch root, vice at the notch root itself.
- The distribution of the two strain energy densities along the minimum cross section not only differ in magnitude, but also in shape.
- The Glinka strain energy density method under estimates the stresses and strains, while the Neuber method overestimates the stresses and strains.
- For the plane stress condition, the two methods appear to give an upper and lower bound. Taking the average of the stresses from each method and determining the strains from this value gave very good results.
- The Glinka method, while under estimating the stress and strain values, produces results two to three times better than the Neuber method for plane strain conditions.

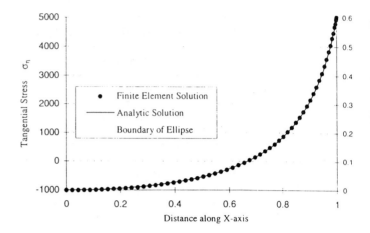

Figure 2. Tangential Stress σ_n Along Elliptical Boundary.

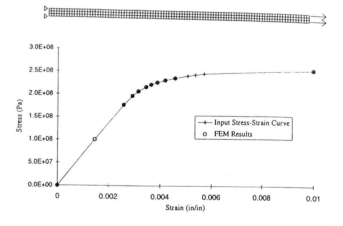

Figure 3. Results of Uniaxial test Case.

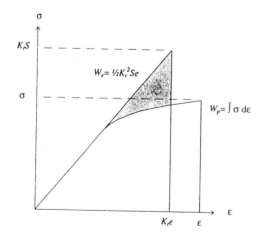

Figure 1. Representation of Strain Energy Density Equivalance Concept

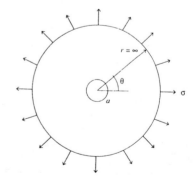

Figure 4. Infinite plate with circular hole

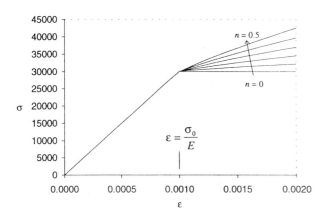

Figure 5. Modified Ludwick Stress-Strain Curve.

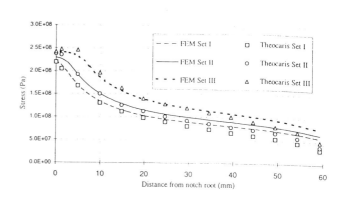

Figure 8. Distribution of σ_y for Theocaris model, load sets I to III.

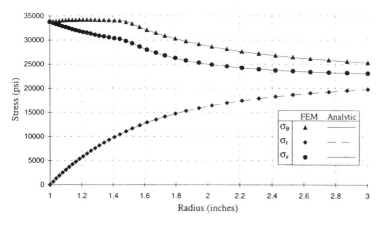

Figure 6. Comparison of FEM stresses with analytic solution for case 2.

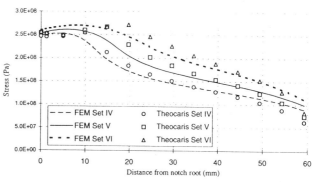

Figure 9. Distribution of σ_y for Theocaris model, load sets IV to VI.

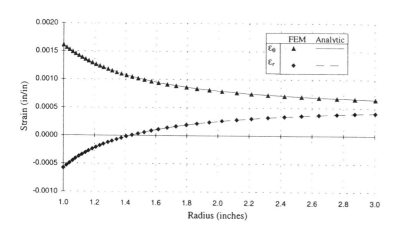

Figure 7. Comparison of FEM strains with analytic solution for case 2.

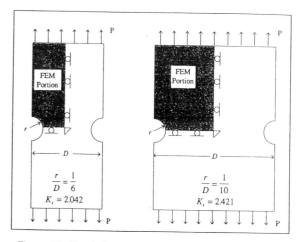

Figure 10. Notch Geometry and Finite Element Boundary Conditions.

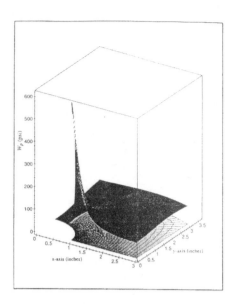

Figure 11. Plastic Strain Energy Density W$_p$ for Narrow Plate in Plane Stress at Load Step 21.

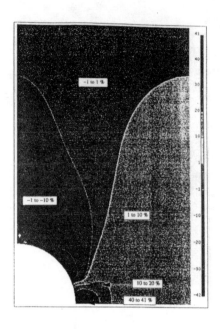

Figure 14. Difference Between W$_p$ and W$_e$ for Narrow Plate in Plane stress at Load Step 21.

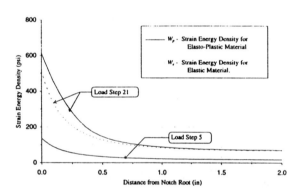

Figure 12. Strain Energy Density along x-asix for Narrow Plate in Plane Stress.

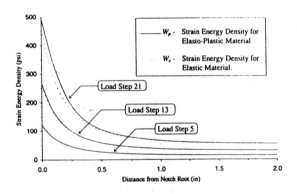

Figure 15. Strain Energy Density along x-axis for Narrow Plate in Plane Strain.

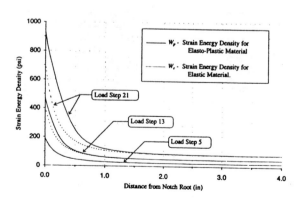

Figure 13. Strain Energy Density along x-axis for Wide Plate in Plane Stress.

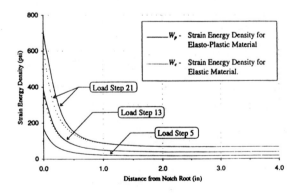

Figure 16. Strain Energy Density along x-axis for Wide Plate in Plane Stress.

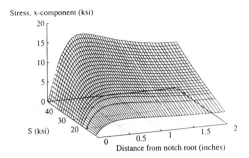

Figure 17a. Distribution of σ_x Along Minimum Cross Section for Narrow Plate in Plane Stress.

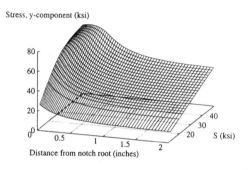

Figure 17b. Distribution of σ_y Along Minimum Cross Section for Narrow Plate in Plane Stress.

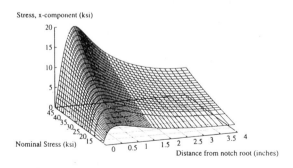

Figure 17c. Distribution of σ_x Along Minimum Cross Section for Wide Plate in Plane Stress.

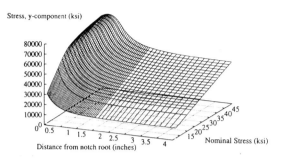

Figure 17d. Distribution of σ_y Along Minimum Cross Section for Wide Plate in Plane Stress.

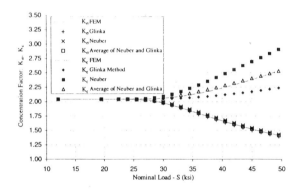

Figure 18a. Stress/Strain Concentration Factors for Narrow Plate in Plane Stress.

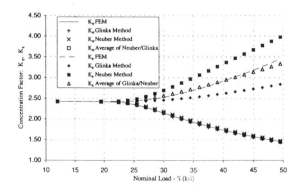

Figure 18b. Stress/Strain Concentration Factors for Narrow Plate in Plane Stress.

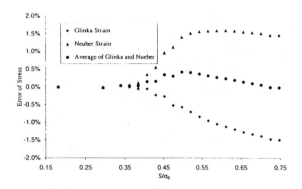

Figure 19a. Error in Notch Stress σ_y for Glinka and Neuber Method, Narrow Plate in Plane Stress.

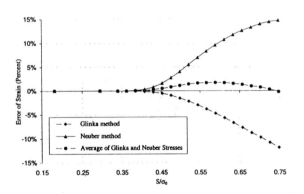

Figure 19b. Error in Notch Strain ε_y for Glinka and Neuber Method, Narrow Plate in Plane Stress.

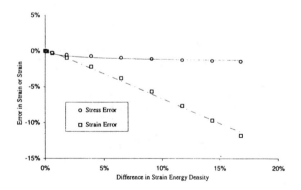

Figure 20. Sensitivity of Stress and Strain Error with respect to Strain Energy DensityError, Narrow Plate in Plane Stress.

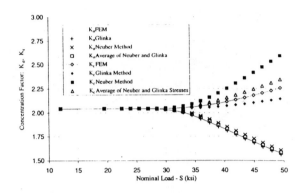

Figure 21a. Stress/Strain Concentration Factors for Narrow Plate in Plane Strain.

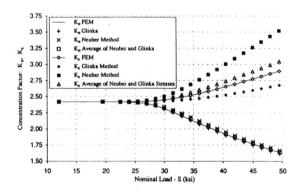

Figure 21b. Stress/Strain Concentration Factors for Wide Plate in Plane Strain.

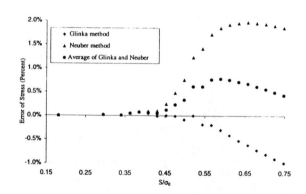

Figure 22a. Error in Notch Stress σ_y for Glinka and Neuber Method, Narrow Plate in Plane Strain.

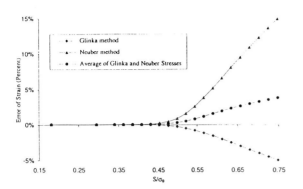

Figure 22b. Error in Notch Strain ε_y for Glinka and Neuber Method, Narrow Plate in Plane Strain.

Material Constant	Case 1	Case 2
σ_0 (ksi)	30,000	30,000
E (ksi)	30,000	30,000
ν	0.3	0.3
σ (ksi)	22.5	22.5
n	0	0.25
ϕ_c	0.3245	0.3245
c	1.5155	1.4789
K	1.3333	1.5020

Table 1 - Material Properties of Infinite Plate with hole

REFERENCES

1. Stephenson, Grant B., "Evaluation of the Strain Energy density Method of Notch Stress Concentration Calculations in the Plastic Range," Master's Thesis, Naval Postgraduate School, Monterey, CA. March 1996.

2. Neuber, H., "Theory of Stress Concentration for Shear-Strained Prismatical Bodies with Arbitrary Nonlinear Stress-Strian Law," *Journal of Applied Mechanics,* 28, 1961, pages 544-550.

3. Molski, Krzysztof and Glinka, Grzegorz, "Method of Elastic-Plastic Stress and Strain Calculation on a Notch Root," *Materials Science and Engineering,* Volume 50, 1981, pages 93-100.

4. Glinka, Grzegorz, "Energy Density Approach to Calculation of Inelastic Strain-Stress Near Notches and Cracks," Engineering Fracture Mechanics, Volume 22, No. 3, 1985, pages 485-508.

5. Glinka, Grzegorz, "Calculation of Inelastic Notch-tip Strain-Stress Histories under Cyclic Loading," *Engineering Fracture Mechanics,* Volume 22, No. 5, 1985, pages 839-854.

6. Glinka, G., Ott, W., and Nowack, H., "Elastoplastic Plane Strain Analysis of Stresses and Strains at the Notch Root," *Journal of Engineering Materials and Technology,* Volume 110, 1988, pages 195 204.

7. Sharpe, W. N. Jr., Yang, C. H., and Trogoning, R. L., "An Evaluation of the Neuber and Glinka Relations for Monotonic Loading," *Journal of Applied Mechanics,* Volume 59, June 1992, pp. S50-S56.

8. Brown, David K., "A Computer Program to Calculate the Elastic Stress and Displacement Fields Around an Elliptical Hole Under Any Applied Plane State of Stress," *Computers and Structures,* Volume 7, No. 4, 1977, pages 571-580.

9. Davis, E. A., "Extension of Iteration Method for determining Strain Distributions to the Uniformly Stresses Plate with a Hole," *Journal of Applied Mechanics,* Volume 30, 1963, page 210-214; discussions, *ibid.,* Volume 31, 1964, pages 362-364.

10. Tuba, I. S., "Elastic-Plastic Stress and Stain Concentration Factors at a Circular Hole in a Uniformly Stressed Infinite Plate," *Journal of Applied Mechanics,* Volume 32, 1965, pages 710-711.

11. Budiansky, Bernard, and Mangasarian, O.L., "Plastic Stress Concentration at a Circular Hole in an Infinite Sheet Subjected to Equal Biaxial Tension," *Journal of Applied Mechanics,"* Volume 27, March 1960, pages 59-64.

12. Chakrabarty, J., *Theory of Plasticity.* McGraw-Hill Book Company, New York, 1987.

13. Durelli, A.J. and Sciammarella, C. A., "Elastoplastic Stress and Strain Distribution in a Finite Plate With a Circular Hole Subject to Unidimensional Load," *Journal of Applied Mechanics,* Volume 30, 1963, pages 115-121.

14. Theocaris, P. S. and Marketos, E., "Elastic-plastic Analysis of Perforated Thin Strips of a Strain-Hardening Material," *Journal of the Mechanics and Physics of Solids,* Volume 12, 1964, pages 377-390.

15. Military Handbook, *Metallic Materials and Elements for Aerospace Vehicle Structures,* Volume 1 of 2 (MIL-HDDBK-5G), Department of Defense, 1 November 1994.

16. Theocaris, P. S. and Marketos, E., "Elastic-plastic Strain and Stress Distribution in Notched Plates Under Plane Stress," *Journal of the Mechanics and Physics of Solids,* Volume 11, 1963, pages 411-428.

17. Dowling, N., Brose, W. R. and Wilson, W. K., "Notched Member Fatigue Life Predictions by the Local Strain Approach," in Fatigue Under Complex Loading, Volume 6 of *Advances in Engineering,* R. M. Wetzel, Ed., Society of Automotive Engineers, Warrendale, PA, 1977.

PVP-Vol. 356, Integrity of Structures and Fluid Systems,
Piping and Pipe Supports, and Pumps and Valves
ASME 1997

INVESTIGATION OF INELASTIC STRESS CONCENTRATIONS
AROUND NOTCHES, PART II: EXPERIMENTAL STUDY

Kenneth S. Long
Young W. Kwon
Gerald H. Lindsey

Naval Postgraduate School
Monterey, California, 93943

ABSTRACT

In Part I of this work, a series of numerical studies were conducted to evaluate the Glinka and Neuber expressions to predict inelastic stress concentrations around notches of specimens under uniaxial tensile loads. Both plane stress and plane strain conditions were investigated. In Part II, experimental studies were conducted to further evaluate Neuber's and Glinka's predictions for inelastic stress/strain concentrations around notches for different loadings and materials. Test materials included an isotropic metal (7075-T6 aluminum alloy) and an orthotropic laminated composite (ARALL-4). In addition, the applied loading condition was an inplane tensile load and an out-of-plane bending load to both materials. Thus, the present study examined the applicability of Glinka's and Neuber's predictions to a laminated composite material as well as to out-of-plane bending load. All these cases had nonuniform stress distributions through the thickness. The results showed that under a tensile load both predictions well bounded the experimental data if a proper stress-strain expression was selected for a given material. However, the predictions for an out-of-plane bending load did not provide the upper and lower bounds of the experimental data. Both theories either under-predicted or over-predicted the experimental inelastic strains, depending on the material.

INTRODUCTION

Part I of the work investigated numerical solutions around notches of tensile specimens under elastoplastic deformation using the finite element technique. Specimens with a circular hole under uniaxial tensile loads with either the plane stress or the plane strain condition were considered in the study. The finite element analysis showed that the elastoplastic strain energy density in the vicinity of a notch tip was greater than the strain energy density obtained assuming elastic only properties. The plane stress condition resulted in a greater difference between the two energy density solutions than the plane strain condition. Because Glinka's predictions of inelastic stresses and strains around notches were based on the elastic strain energy density, his solutions underestimated the results of the finite element solutions. On the other hand, Neuber's prediction was always overpredicted the elastoplastic stresses and strains around notches. As a result, a modified technique based on averaging Glinka's and Neuber's solutions was proposed in Part I, and the modified solutions greatly improved the accuracy of the predicted elastoplastic stresses and strains around notches.

The studies in Part I analyzed only the tensile specimens of an isotropic material. Therefore, Part II of the work was to investigate the inelastic solutions around notches of specimens made of an isotropic metal (7075-T6 aluminum alloy) and an orthotropic laminated composite (ARALL-4). The specimens were subject to inplane tensile loads and out-of-plane bending loads, respectively. An experimental study was conducted in Part II, and the experimental solutions were compared to the solutions predicted from Glinka's and Neuber's expressions. Thus, the objective of this study was to evaluate both Glinka's and Neuber's predictions of inelastic stresses and strains around notches when the specimens had nonuniform stresses through the thickness. That is, the laminated composite specimen under an inplane tension had both elastic and elastoplastic deformations through the thickness because some laminae had elastic deformation and others had elastoplastic deformation. Further, an out-of-plane bending load produced nonuniform stresses through the thickness regardless of the material used.

The outline of this paper is as follows: In the next section, a description of preliminary testing performed on 7075-T6 aluminum alloy and ARALL-4 composite to determine their elastic and plastic material properties under uniaxial tension as well as under four-point bending is given. The following section describes the tensile tests of perforated specimens made of the two materials, Glinka's and Neuber's theories, and experimental results and discussion. The subsequent section discusses the flexural tests using four-point bending and the results. The experimetal results were compared to the analytically predicted solutions. Finally, conclusions follow at the end.

TESTING FOR MATERIAL PROPERTIES

This section describes the preliminary tests and results that were conducted on two types of material: 7075-T6 aluminum alloy and Aramid Aluminum Laminate (ARALL-4). Preliminary tests were required to verify and establish properties of the materials because that information was required for tests described in subsequent studies.

7075-T6 Aluminum Alloy

The aluminum alloy used in this study was cut from one sheet of 7075-T6 aluminum. This aluminum sheet metal was also identified by the Government Mill Form and Condition designation as QQ-A-250-12. Initial dimensions of the large sheet were 54 in.(137.2 cm) x 46 in.(116.8 cm) x 0.125 in.(0.318 cm).

Tension tests were conducted on several test coupons to verify the 7075-T6 aluminum sheet metal material properties used in this study as compared with published 7075-T6 aluminum material properties. The test coupon design and the testing methods used were governed by American Society for Testing of Materials standard test methods [1995a]. Figure 1 shows the dimensions of the tension test coupons used for material verification. The methods listed in [ASTM 1995a] were used to determine the material properties of yield strength (offset=0.2%) and ultimate tensile strength. The testing methods used to determine Young's modu-

lus and Poisson's ratio were listed in [ASTM 1988] and [ASTM 1992a], respectively.

Test coupon width and thickness were measured using a Mitutoyo Digital Micrometer, accurate to 0.00005 in.(0.001 mm). The 90 degree strain gage rosettes used were Measurements Group Inc. type CEA-13-125UT, 120 ohm, gage factor 2.085±0.5% (longitudinal) and 2.115±0.5% (transverse). The Instron dynamic extensometer used had a 3.5 in.(8.9 cm) gage length and ±0.2 in.(0.51 cm) range. The Instron Model 4507 tension/compression test machine was used with a 200 kN load cell. Load and displacement data were obtained using the Instron Model 4500 data acquisition tower with the Instron series IX automated materials testing software, version 5.28. Strain gage readings were acquired through the Measurements Group P-3500 strain indicator/SB-100 switch and balance unit combination. These two units were used to acquire all the strain gage readings in following tests.

Table 1 shows the comparison of the experimentally determined material properties with published material properties [ASM 1985] for 7075-T6 aluminum alloy. The experimental results provided in Table 1 were based on data and calculations from three test coupons. Three test coupons were cut in the longitudinal orientation from the aluminum sheet. Because transverse mechanical properties of many products, particularly tensile strengths and ductility, can be less than those in the longitudinal direction [ASM 1985], test coupons were cut from the sheet in the transverse direction to verify isotropic behavior of the aluminum sheet. The results showed almost identical material properties in both longitudinal and transverse directions, with the largest difference being less than two percent. This confirmed the isotropic material behavior of the aluminum alloy.

Elastic Modulus For elastic modulus calculations, each tensile test coupon was tested three times in the elastic range by cyclic loading, as per Reference [ASTM 1988]. The modulus values calculated for each of the three runs were averaged, resulting in an average modulus value for that test coupon. The elastic modulus used in this study was the average modulus value of the three test coupons oriented in the longitudinal direction. During the tensile tests, strain data was collected from two sources. One source was a strain gage rosette mounted on the geometric center of the test coupon. The other source was an extensometer. Elastic modulii were calculated using stress and both strain data sources. As before, modulus values calculated for each run were averaged to result in a modulus value for each test coupon. Averaging the longitudinally oriented test coupons' modulii based on extensometer data resulted in Young's modulus of 10,269 ksi (70.6 GPa). The average longitudinal Young's modulus based on strain gage data was 10,121 ksi (69.6 GPa), a 1.5% difference. Reference [ASTM 1992] describes the standard method of recording strain data with the use of extensometers. In order to prevent damage to the extensometers by subjecting them to impact forces that occur at ultimate failure, they were removed just prior to sample failure. The only means of obtaining strain data up to ultimate failure was with the use of strain gages.

Poisson's Ratio Reference [ASTM 1992a] provided the standard method of experimentally determining Poisson's ratio using extensometers. Since the use of strain gages and extensometers produced similar results for Young's modulus, strain gage based data was also used for determining Poisson's ratio.

Nonlinear Coefficients A relation most often used for strain in the nonlinear region is the Ramberg-Osgood relation [1943]

$$\epsilon = \frac{\sigma}{E} + \left(\frac{\sigma}{K}\right)^{1/n} \qquad (1)$$

where ϵ and σ are the strain and stress, respectively, and E is the elastic modulus. Furthermore, K is the strength coefficient and n is the strain-hardening index. K and n were experimentally determined from the results of the tension tests. For most metals, a log-log plot of true stress versus true plastic strain is modeled as a straight line [Bannantine, Comer, and Handrock 1990]. However, the relationship often deviates from linearity at low strains and/or high strains [ASM 1992]. The slope of the line is n and the intercept with the log true stress axis is K. Figure 2 shows the plot of log true stress versus log true plastic strain for an aluminum sample. K and n values were calculated for every sample and averaged. It is obvious that the data is not linear throughout the plastic range, but it is satisfactory to calculate K and n for the strain range over which the log-log plot is linear. Therein lies the question of which region of the plot should be considered linear. Two cases of linear approximations were investigated. Case 1 considers the data points of log true plastic strain between -4 and -2.4. Case 2 considers the data points of log true plastic strain between -3.2 and -2.4. A least squares approximation of the data for both cases is shown in Figure 2. The results were $K = 110.13$ ksi and $n = 0.0790$ for case 1, and $K = 114.88$ ksi and $n = 0.0946$ for case 2. It will be shown later that the case 2 linear approximation for K and n will provide slightly better results. Thus, the average K and n values used

in this study are 114.88 ksi and 0.0946, respectively.

Statistical Analysis A statistical study was employed to determine the errors in the least squares method of approximation to Young's modulus calculations. Correlation coefficients (r) were calculated for each run and averaged for each test coupon. The test coupons' averaged coefficients were then averaged to obtain a single correlation coefficient. The average correlation coefficient for longitudinally oriented stress-strain plots in the elastic region was 0.99979. When the data is basically linear, as was the case in these tensile tests, the correlation coefficient is not a very sensitive indicator of the precision of the data. McClintock [1987] found that $(1 - r^2)^{1/2}$ is a better indicator of the curve fit quality. Here, the corresponding value is 0.0146. This value indicates that the vertical standard deviation of the data is only 1.46 percent of the total vertical variation caused by the straight-line relationship between stress and strain [Beckwith, Marangoni, and Lienhard, 1993].

Aramid Aluminum Laminate

An aramid aluminum laminate (ARALL) has also been considered in this study. ARALL is classified as a polymer matrix composite material. The specific fiber-reinforced composite used was ARALL-4, which has a 5/4 configuration. There were five layers of 2024-T8 aluminum alloy and four layers of unidirectional and continuous Kevlar 49 aramid fibers. The laminate was symmetric and balanced with respect to the test direction. The 2024-T8 aluminum laminae were isotropic and all the aramid laminae were oriented in the same direction. Two sheets of the composite, measuring 12 inches by 12 inches each, were used for testing.

Four samples were initially cut from the sheets provided. Two samples were machine finished to the same dimensions as the 7075-T6 aluminum samples used for the material verification tensile tests (see Figure 1). One of these samples had the fibers oriented parallel to the loading direction and the other had the fibers oriented transverse to the loading direction. This design allowed for the calculation of longitudinal and transverse modulii. The other two samples were machine finished to a rectangular shape of 1 in.(2.54 cm) by 9 in.(22.86 cm). These samples had fibers oriented only in the longitudinal direction and were used in flexural tests to determine the bending modulus. The equipment used for the laminate material property tensile tests was the same as that used for the 7075-T6 aluminum material property tests. For the flexural tests, a four-point bending apparatus was constructed to determine bending modulus, and it is shown in Figure 3.

Four-point loading at the quarter points was the load configuration, which is a typical loading condition for high-modulus materials. The sample length was 7 1/4 in.(188 mm) and laminate thickness was 3/32 in.(2.4 mm). Sample length was based on an ASTM recommended 60:1 span-to-thickness ratio for high-modulus composites. Deflection was measured using a Laboratory Devices Co. dial gage, accurate to 0.001 in. The strain gages used were Measurements Group Inc. type CEA-13-250UN, 350 ohm, gage factor 2.12±0.5%. For both tensile and flexural tests, the Instron test machine was used in the manual load control mode to record load, strain and deflection data.

The reference used for the tensile test to determine Young's modulus and Poisson's ratio was ASTM 3039 [1995b] and the reference used for the flexural test to determine the bending modulus was ASTM 790 [1992b].

Elastic Modulus Two samples were tested to determine the smeared elastic modulus. This was the modulus for the laminate rather than a modulus for each of the different lamina. One sample was used to calculate E_1 (Young's modulus in the fiber direction) and the other was used to calculate E_2 (Young's modulus transverse to the fiber direction). Two strain gages were placed on each sample. One gage was oriented in the fiber direction and the other gage was oriented transverse to the fiber direction. Although each sample was tested to ultimate failure, only data in the linear elastic range was used to calculate modulus values. For the sample with fibers oriented longitudinal to the loading direction, Figure 4 shows the stress-strain curves in the longitudinal and transverse directions. Absolute values of the transverse strains were plotted in the figure. A similar plot was obtained for the stress and strain in the longitudinal and transverse directions when a sample was oriented transverse to the loading direction. The best fit approximation of the slope of the data in the elastic region was the modulus value. For the longitudinal modulus, $E_1 = 9{,}100$ ksi (62.6 GPa) and for the transverse modulus, $E_2 = 6{,}600$ ksi (45.4 GPa). These modulii show that a sample with fibers oriented longitudinal to loading direction is stiffer than a sample with fibers oriented transverse to loading direction. A comparison of the calculated E_1 value to published elastic modulii (tension) for 2024 aluminum and Kevlar 49 fibers was made. Because elastic modulii are dependent on laminate thickness and lamina orientation, a standard elastic modulus for the ARALL4 material used was not available. The published elastic modulus (tension) for 2024 aluminum is 10,500 ksi (72.4 GPa) [ASM 1985] and for Kevlar 49 fibers is 18,100 ksi (125 GPa) [Chawla 1987]. Although the elastic modulus for

Kevlar 49 is high, the lamina did not consist of a 100 percent volume fraction of Kevlar 49 fibers. These lamina contained an undetermined amount of resin, interface bonding material, and volume fraction of Kevlar 49 fibers much less than 100 percent. The low volume fraction of Kevlar 49 fibers effectively lowers the lamina elastic modulus to a percentage equal to the volume fraction. Thus, laminate elastic modulus values greatly depend on the volume fraction of the fibers within the lamina.

Bending Modulus An alternative approach to tensile testing of unidirectional polymeric-matrix composites involves the bending of beam-type samples [ASM 1992]. The flexural test method was used to determine a smeared bending modulus for the overall laminate vice individual 2024 aluminum or aramid lamina modulii. Two samples were also used in flexural tests to determine the bending modulus, E_B. The purpose of using two samples was to achieve repeatability of the data. Since the results for each sample were nearly identical, only one sample's data was presented. Figure 5 shows the plot of load versus deflection in the elastic range. For the test, $E_B = 11{,}200$ ksi (77.0 GPa). The bending modulus was calculated using

$$E_B = \frac{0.17 L^3 m}{b t^3} \qquad (2)$$

where L = length of support span, b = width of beam, t = thickness of beam, and m = slope of load-deflection curve in elastic range. Equation (2) came from ASTM D790-92 (1992b).

For laminate composites, the tensile modulus and bending modulus are expected to be different. With highly anisotropic laminates, the bending modulus is strongly dependent on the lamina-stacking sequence [ASTM 1992b]. A tensile modulus is not stacking-sequence dependent. In the flexural test conducted, the bending modulus E_B was 19 percent higher than the longitudinal modulus E_1. This can be explained by the different stiffnesses of the lamina. The 2024 aluminum lamina are stiffer than the aramid fiber lamina. The outermost laminae of the composite were the 2024 aluminum lamina. The stiffer outermost laminae results in higher bending stiffness but has no effect on the tensile stiffness.

Poisson's Ratio The strain data taken from each sample used in the elastic modulus section was used to determine Poisson's ratio. Two Poisson's ratios were calculated. ν_{12} corresponds to the sample with fibers oriented longitudinally to the loading direction, while ν_{21} is for the sample with fibers oriented transverse to the loading direction. Figure 6 shows Poisson's ratio, ν_{12}, plotted versus load in the elastic region. The average value is drawn by the horizontal line and calculated by

$$\nu_{avg} = \frac{1}{n} \sum_{i=1}^{n} \left| \frac{\epsilon_{ti}}{\epsilon_{li}} \right| \qquad (3)$$

where ϵ_{li} is the longitudinal strain, ϵ_{ti} is the transverse strain, and n is the total number of data points in the elastic region. This calculation was used to find $\nu_{12} = 0.323$. The relation

$$\frac{\nu_{12}}{E_1} = \frac{\nu_{21}}{E_2} \qquad (4)$$

was used to predict ν_{21}, since the other values were known. The calculation yielded $\nu_{21} = 0.234$.

TENSION TEST

Tension tests on circular notched samples were conducted for both the 7075-T6 aluminum and the aramid aluminum laminate samples. The purpose of the tests on the 7075-T6 aluminum was to validate Glinka's and Neuber's predictions of inelastic stress concentrations around notches. Some experimental validation for a tensile specimen with a circular hole was conducted in a previous study [Glinka 1985]. However, the study was for an isotropic, homogeneous metallic material. Thus, the present tests included both aluminum and ALLAL-4 composite specimens.

This section discusses the equipment apparatus used for tension testing, Glinka's and Neuber's theories, the method of computations for each material, and the results of the comparison of Glinka's and Neuber's theoretical inelastic stress-strain approximations to experimental stresses and strains. Test results indicated that the Glinka and Neuber methods based on the Ramberg-Osgood stress-strain curve worked well as expected for the 7075-T6 aluminum sample but not for the ARALL 4 sample. Good theoretical inelastic strain predictions were made using Glinka and Neuber methods based on the bilinear stress-strain equation.

Experimental Apparatus

Figure 7 shows the geometry and dimensions of the circular notched samples used for the tension tests. The length of the 7075-T6 aluminum sample was 15 in. and the length of the ARALL-4 sample was 12 in. Both sample types had machine finished edges. The circular notch was milled through in the 7075-T6 aluminum sample and drilled through in the ARALL sample. The circular notch was located in the geometric center of the sample. Also shown in Figure 7 is the location

and orientation of the strain gages. The 10-element strain gages used were Measurements Group Inc. type CEA-09-020PF, 120 ohm, gage factor 2.00±3.0%. The single element strain gages used were Measurements Group Inc. type CEA-13-032UW, 120 ohm, gage factor 2.11±1.0%. The two single element strain gages were positioned back-to-back only on the 7075-T6 aluminum sample. Having the two single gages positioned back-to-back and midway between the outer edge of the hole and edge of the sample permitted the comparison of strains on both sides of the sample. Greater strains on one side of the sample would indicate the presence of bending stresses caused by misaligned sample gripping. Tension tests were conducted on the Material Test System (MTS) Model 810 tension/compression test machine. The load cell cartridge was rated at 25 kips and the maximum deflection cartridge was rated at 1 inch. The MTS test machine was selected for testing because the grip width was 2 in., which was wide enough to extend across the entire width of the circular notched samples. For the 7075-T6 aluminum sample, the length of the sample gripped by the MTS machine was 2 in. on each end. The ARALL-4 sample gripped length was 1.25 in. on each end. The MTS machine was manually load controlled. Using a 10^{-4} in./sec loading speed, the loading was periodically paused to allow recording of load and strain data. Samples were loaded to ultimate failure.

Glinka's and Neuber's Theories
Ramberg-Osgood Stress-Strain Relation

In Part I of this paper, both Glinka's and Neuber's equations for predicting inelastic stresses-strains near notches were presented in detail. Thus, a short development of the theories is presented below as much as needed to discuss the results. Both theoretical predictions were based on the Ramberg-Osgood stress-strain expression which was given in Eq. (1). The strain energy density distribution ahead of a notch tip, W_e, can be calculated as

$$W_e = \int_0^{e_{ij}} s_{ij} de_{ij} \qquad (5)$$

where s_{ij} is the local elastic stress tensor and e_{ij} is the local elastic strain tensor. For localized small-scale plastic yielding, the energy density in the plastic zone is nearly equal to that in the elastic zone. Thus, the energy density W_σ in the plastic zone is equal to that calculated on the basis of an elastic solution

$$W_e = W_\sigma \quad \text{or} \quad \int_0^{e_{ij}} s_{ij} de_{ij} = \int_0^{\epsilon_{ij}} \sigma_{ij} d\epsilon_{ij} \qquad (6)$$

where σ_{ij} is the local inelastic-plastic stress tensor and ϵ_{ij} is the local elastic-plastic strain tensor.

The elastic stress at the notch tip is

$$\sigma_y = K_t S_n \qquad (7)$$

where S_y is the local elastic stress component in the y direction (i.e. the loading direction), K_t is the stress concentration factor, and S_n is the nominal stress. For uniaxial stress at the notch tip, Eq. (5) can be written in the form

$$\frac{(K_t S_n)^2}{2E} = \frac{\sigma_y^2}{2E} \qquad (8)$$

This states that elastic strain energy density W_e at the notch tip is equal to the product of the strain energy density W_s due to the nominal stress S_n and the square of the stress concentration factor K_t.

$$K_t^2 W_s = W_\sigma \qquad (9)$$

With localized yielding at the notch tip, the energy density should be calculated with respect to the non-linear stress-strain relation (1), then Eq. (9) takes the form

$$K_t^2 \frac{S_n^2}{2E} = \frac{\sigma_y^2}{2E} + \frac{\sigma_y}{n+1} \left(\frac{\sigma_y}{K} \right)^{1/n} \qquad (10)$$

When the nominal stress S_n has exceeded the proportional limit, equation (10) takes the form

$$K_t^2 \left[\frac{S_n^2}{2E} + \frac{S_n}{n+1} \left(\frac{S_n}{K} \right)^{1/n} \right] = \frac{\sigma_y^2}{2E} + \frac{\sigma_y}{n+1} \left(\frac{\sigma_y}{K} \right)^{1/n} \qquad (11)$$

Equations (10) and (11) enable one to calculate the inelastic stress σ_y and strain ϵ_y at the notch tip for a given nominal stress S_n and stress concentration factor K_t. S_n is easily calculated from

$$S_n = \frac{P}{A_{min}} \qquad (12)$$

where P is the tensile load and A_{min} is the minimum cross-sectional area of the sample. All these equations were based on Glinka's assumption.

Neuber [1961] derived a relation to calculate local inelastic strain ϵ_y and stress σ_y. In the elastic region, the energy-density method and Neuber's rule are the same. However, in the inelastic region, Neuber's expression takes the form

$$K_t^2 \left[\frac{S_n^2}{2E} + \frac{S_n}{2} \left(\frac{S_n}{K} \right)^{1/n} \right] = \frac{\sigma_y^2}{2E} + \frac{\sigma_y}{2} \left(\frac{\sigma_y}{K} \right)^{1/n} \qquad (13)$$

but the right hand side does not represent energy density. σ_y is the Neuber determined inelastic stress.

Bilinear Relation As shown later in this section, the predicted theoretical inelastic notch strains for the ARALL-4 composite were unsatisfactory when the Glinka and Neuber methods were used as developed in the previous paragraphs. One major reason was that the Ramberg-Osgood stress-strain relation was not suitable for the ARALL-4 composite. Therefore, the Glinka and Neuber equations were rederived using the bilinear stress-strain behavior of the ARALL-4 composite. In the upper linear region of the bilinear stress-strain diagram, Eq. (10) becomes

$$K_t^2 \frac{S_n^2}{2E} = \frac{\sigma_o^2}{2E} + \frac{(\sigma_y^2 - \sigma_o^2)}{2E_t} \quad \text{for } \sigma_y > \sigma_o \qquad (14)$$

where E_t is the modulus in the upper linear region and σ_o is the stress at the intersection of two linear regions. Theoretical inelastic notch strains for the bilinear Glinka method are obtained using

$$\epsilon = \frac{\sigma_y - \sigma_o}{E_t} + \epsilon_o \qquad (15)$$

where ϵ_o is the strain at the intersection of two linear regions.

Neuber's method may be redeveloped to account for the bilinear stress-strain behavior, such that

$$K_t^2 \frac{S_n^2}{2E} = \frac{\sigma_y}{2} \left(\frac{\sigma_y - \sigma_o}{E_t} + \frac{\sigma_o}{E} \right) \quad \text{for } \sigma_y > \sigma_o \qquad (16)$$

where σ_y represents the stress from Neuber's method. Theoretical inelastic notch strains for the bilinear Neuber's method are also found using (15).

Method of Computations

7075-T6 Aluminum Alloy
For notch stresses lower than the yield stress, theoretical elastic strains were calculated using Hooke's Law. Equations (11) and (13) were used to compute Glinka's and Neuber's theoretical inelastic stresses based on the nominal stresses calculated in Eq. (12). Because Eqs (11) and (13) are nonlinear, the bisection numerical technique was employed to solve for the theoretical inelastic stresses. Theoretical inelastic strains were calculated by substituting the theoretical inelastic stresses from Eqs (11) and (13) into Eq. (1).

ARALL-4 Composite
In the elastic region, strains were calculated using Hooke's Law. Inelastic strains were calculated using Eqs (11), (13) and (1). These equations produced unsatisfactory results. Better results were obtained with Glinka's and Neuber's methods, Eqs (14) and (16). Inelastic stresses were calculated from Eqs (14) and (16) and substituted into Eq. (15) to obtain inelastic strains.

RESULTS

7075-T6 Aluminum Alloy Figure 8 shows that the Ramberg-Osgood relation produces a good stress-strain approximation to experimental data for 7075-T6 aluminum. However, the decision as to which data points to consider to determine the strength coefficient K and strain hardening index n can affect the results. Figure 9 shows the comparison of Glinka's theoretical notch strain to experimental notch strain for tensile loading. The experimentally determined stress-strain curve is added to Figure 9 for reference. Two theoretical approximations, cases 1 and 2, are shown. In Case 1, $K = 110.13$ ksi and $n = 0.0790$. In Case 2, $K = 114.88$ ksi and $n = 0.0946$. As mentioned in the previous section and which is now apparent, case 2 K and n material property values provided a closer approximation to experimental results. During elastic behavior, the experimental, theoretical, and stress-strain curves are the same. The deviation of the experimental and theoretical curves from the stress-strain curve occurs at the beginning of inelastic behavior. In the inelastic range, the results confirm Glinka's results [1985] in that, "The difference between calculated and measured notch strains were smaller than 10 percent ...". For the 7075-T6 aluminum sample, the maximum calculated and measured strain difference was 11.8 percent in the inelastic region. This experiment also confirmed the Lee et al. [1995] claim that Glinka's rule underestimates the notch strain in the high yield deformation region and overestimates the local strain in the low plasticity zone. Figure 10 shows the comparison of Glinka and Neuber methods for determining theoretical circular notch strain in 7075-T6 aluminum. As Glinka [1985] indicated, Neuber's rule underestimates notch strain. However, Sharpe et al. [1992] reported that the Neuber rule is the best single model for predicting notch strains in a plane stress condition. This is because the Neuber rule is more conservative than the Glinka rule. For large plastic deformation regions, the Glinka plane stress model gives an upper bound of strain. The sample reached ultimate failure at a tensile load of 13,270 lbs; the corresponding nominal stress was 85.0 ksi.

The plot of theoretical and experimental circular notch strains proved to be very sensitive to the selection of the elastic stress concentration factor K_t. The elastic stress concentration factor gradient ahead of the notch was calculated using Howland's method of coefficients [1930]. The ratio of hole diameter to plate width was $\lambda = 0.334$. Howland only calculated coefficients for the ratios of $\lambda = 0.1, 0.2, 0.3, 0.4$ and 0.5. Thus, the coefficients used in this study had to be interpolated from Howland's table of coefficients. The stress

concentration factor gradient away from the notch is shown in Figure 11. The gradient is steepest near the notch and decreases away from the notch. Superimposed on the gradient curve are the locations of individual strain gages (indicated by horizontal error bars) from the 10-element strip gage. The average stress concentration factor over the strain gage range is indicated by a circle. Next to the hole, $K_t = 2.26$ using Howland's method. Integrating the gradient over the range of the first strain gage's measuring range to find the average, $K_t = 2.04$. The stress concentration factor used for the 7075-T6 aluminum tension test results was $K_t = 1.90$. Although this K_t fell within the range of the first strain gage, it was slightly lower than the average because the first gage was located a small distance away from the edge of the notch. Strain gradients ahead of the circular notch were examined for several nominal stresses. At low nominal stresses, the gradient was nearly linear. However, as stress was increased, the gradient next to the edge of the notch greatly increased.

ARALL-4 Composite Both Glinka's equation (11) and Neuber's equation (13), that were used to predict theoretical inelastic stress, depended on the strength coefficient, K, and the strain hardening index, n. Assuming the stress-strain behavior for the ARALL-4 composite could be represented by the Ramberg-Osgood equation (1), K and n were calculated and applied to Eqs (11) and (13). The calculated values were $K = 889$ ksi and $n = 0.570$. When these values were used for Eqs (11) and (13), Glinka's method greatly overestimated the inelastic notch strains. Neuber's method also produced nearly the same overestimation for inelastic strains that Glinka's method did.

One of the main reasons was that the Ramberg-Osgood equation did not model bilinear stress-strain behavior of the ARALL-4 composite. Consequently, the Glinka and Neuber equations produced unsatisfactory results. The prediction for theoretical inelastic notch strains can be greatly improved by using the bilinear Glinka and Neuber equations, i.e. Eqs (14) and (16). The comparison of the bilinear Glinka and Neuber methods for the ARALL-4 composite in tension is shown in Figure 12. These two equations based on the bilinear stress-strain curve gave a bound to the experimental data. The overestimation in inelastic strains using the bilinear Neuber method provides a more conservative approximation. An underestimation in notch strains resulted from the bilinear Glinka method.

FLEXURAL TEST

Flexural tests were conducted on both 7075-T6 aluminum and ARALL-4 samples. Four-point loading was applied to the quarter points of the samples. This created out-of-plane bending stresses around the circular notch. Flexural testing was conducted to determine whether the Glinka and Neuber relations established for tensile loading could be used to predict local inelastic strains for bending. This section discusses the experimental apparatus used in conducting the tests, the flexural test procedure, the theory used to determine inelastic bending stresses for both materials, and the test results for the 7075-T6 aluminum and ARALL-4 composite.

Experimental Apparatus

The design of the circular notched samples used in the flexural tests were the same as that used in the tension tests. Sample sizes and dimensions listed in Figure 7 also apply to flexural test samples. The same 10-element strip gages were used but the single back-to-back strain gages were omitted. Also, the Instron test machine used for material property verification was used for flexural tests. However, in order to take incremental load and strain data, the Instron was operated in the manual load control mode. A four-point bend test apparatus was manufactured in accordance with ASTM specifications [1992b] and the dimensions are shown in Figure 3.

Bending Test procedure

A sample was placed on the base of the beam test apparatus and symmetrically aligned. Each strain gage of the 10-element strip gage was zeroed. The top of the beam test apparatus, which weighed 9.6 lbs, was placed on the sample and symmetrically aligned. This condition accounted for the first data point. Incremental loads were then applied to the sample to obtain the remaining data points. For the 7075-T6 aluminum sample, the test was concluded when the sample deflected approximately 2.1 inches, almost touching the base of the beam test apparatus. At this point, the corresponding load was 365 lbs. The maximum load reached before the ARALL-4 sample deflected into the base of the bend test apparatus was 410 lbs. Because the ARALL4 sample were shorter in length, it could withstand a greater load.

Theory

As tensile loading is increased for a circular notched sample made of an isotropic, homogeneous material, plastic yielding first occurs at the edge of the notch and progresses away from the notch, towards the edge of the sample. At any location away from the notch, the amount of plastic yielding at that distance will be the same through the sample thickness. Just as in tensile

loading, the plastic yielding for bending begins at the edge of the circular notch and progresses outward. However, at a given location away from the notch, as loading is increases plastic yielding begins on the surface of the sample and progresses inward, through the thickness, towards the neutral axis. A small region of plastic yielding is often referred to as localized plastic yielding.

7075-T6 Aluminum Alloy The same Glinka relation (10) presented in the previous section for localized yielding at the notch tip under tensile loading can be used for bending. The Neuber relation for bending is similar to Eq. (10) but the $n + 1$ in the second term on the right hand side is replaced with 2. Here, nominal bending stress is calculated by

$$S_n = \frac{6M}{t^2(D - 2d)} \tag{17}$$

where M is the bending moment, t is the sample thickness, D is the plate width and d is the notch diameter.

ARALL4 Composite The bilinear Glinka and Neuber equations, Eqs (14) and (16), presented in the previous section can be also used for bending with the nominal stress from Eq. (17).

RESULTS
7075-T6 Aluminum Alloy

Figure 13 shows the comparison of Glinka and Neuber methods for determining theoretical inelastic notch strain in out-of-plane bending. The Glinka approximation was much closer to the experimental strains than the Neuber approximation. For low and moderate local inelastic strains, the Glinka method overestimated the strain, but in the high yield region the Glinka method underestimated the local inelastic strain.

Figure 14 shows several strain gradients ahead of the circular notch. These gradients are slightly different when compared to those for the 7075-T6 aluminum samples in tension loading. The slope of the gradient next to the notch was lower for bending than tension. As in the tension test, the strain gradients in bending approached a constant value away from the notch.

The stress concentration factor used in Figures 13 and 14 was 1.20. Peterson [1974] has plotted several curves for stress concentration factors subject to transverse bending of a finite-width plate with a circular hole. The experimental circular notch diameter (d) to sample width (D) ratio was 0.34. Peterson plotted the stress concentration factor curves for d/D from 0 to 0.2. If the appropriate curve is extrapolated to $d/D = 0.34$, the resulting stress concentration factor next to

46

the notch is approximately 1.40. As discussed in the previous section, the strain gage was located a small distance from the edge of the notch. The strain gage averaged the stress concentration factor gradient over the gage's measured area. The 15 percent difference in experimental and theoretical stress concentration factors can also be attributed to a slight off-centering of the circular notch by 0.03 in.(0.76 mm) during the machining process.

ARALL-4 Composite

As discussed in the tensile test section, the Glinka and Neuber equations associated with the Ramberg-Osgood stress-strain curve greatly overestimated the inelastic notch strains for the ARALL-4 composite under out-of-plane bending. Figure 15 shows the comparison of the bilinear Glinka and Neuber methods for out-of-plane bending. The Glinka and Neuber methods based on the bilinear stress-strain curve greatly improved the inelastic strain approximation over those based on the Ramberg-Osgood curve. However, the two methods underestimated experimental notch strains at large plastic deformations. For small plastic deformations, the bilinear Neuber method was a slightly conservative approximation. Figure 16 shows several strain gradients ahead of the circular notch for the ARALL-4 composite in out-of-plane bending. The stress concentration factor used in the ARALL-4 tests was 1.05.

CONCLUSIONS

Because the behavior of inelastic stresses/strains around a circular notch is complex, approximate methods have been proposed by Glinka and Neuber. The present study examined the methods for cases where stresses were nonuniform through the specimen's thickness like out-of-plane bending loads and for laminated composite specimens. Specific findings of this study include:

Neuber's method always predicted inelastic circular notch strains that were larger than those using Glinka's method.

While the inelastic stress-strain curve for the 7075-T6 aluminum alloy could be well represented by the Ramberg-Osgood equation, the bilinear stress-strain equation was proper for the ARALL-4 composite. As a result, predictions of inelastic notch strains were made based on proper stress-strain curves for both materials.

For a uniaxial tensile load, Neuber's prediction of inelastic notch strains in general were overestimated while Glinka's prediction was underestimated compared to the experimental data for both materials. Thus, the averaging technique proposed in Part I of this study would work well for both materials.

When specimens were under out-of-plane bending loads, Neuber's and Glinka's predictions of notch strains did not provide proper bounds of the experimental results. For the aluminum alloy, both theories overpredicted the test data and Glinka's result was closer to the experimental result than Neuber's prediction. On the other hand, for the ARALL-4 composite, both theories underpredicted at large strains. Thus, the Neuber theory was not conservative. Overall, the Neuber prediction was better than the Glinka prediction.

REFERENCES

ASM (American Society for Metals) (1985), *Metals Handbook*, ASM International, Vol. 2.

ASM (American Society for Metals) (1992), *Tensile Testing*, ASM International.

ASTM E111-82 (1988): "Standard Test Method for Young's Modulus, Tangent Modulus, and Chord Modulus", American Society for Testing of Materials, ASTM Standards.

ASTM E132-86 (1992a): "Standard Test Method for Poisson's Ratio at Room Temperature", American Society for Testing of Materials, ASTM Standards.

ASTM D790-92 (1992b): "Standard Test Methods for Flexural Properties of Unreinforced and Reinforced Plastics and Electrical Insulating Materials", American Society for Testing of Materials, ASTM Standards.

ASTM E8-95a (1995a): "Standard Test Methods for Tension Testing of Metallic Materials", American Society for Testing of Materials, ASTM Standards.

ASTM D3039-95 (1995b): "Standard Test Method for Tensile Properties of Polymer Matrix Composite Materials", American Society for Testing of Materials, ASTM Standards.

Bannantine, J., Comer, J., and Handrock, J. (1990), *Fundamentals of Metal Fatigue Analysis*, Prentice Hall, p. 45.

Beckwith, T., Marangoni, R., and Lienhard, V. (1993), *Mechanical Measurements*, Addison-Wesley.

Chawla, K. K. (1987), *Composite Materials: Science and Engineering*, Springer-Verlag, Table 2.5, p. 37.

Glinka, G. (1985), "Energy Density Approach to Calculation of Inelastic Strain-Stress Near Notches and Cracks", *Engineering Fracture Mechanics*, Vol. 22, No. 3, pp. 485-508.

Howland, R. C. J. (1930), "On the Stresses in the Neighbourhood of a Circular Hole in a Strip Under Tension", *Philosophical Transactions of the Royal Society of London*, Series A229, pp. 49-86.

Lee, Y., Chang, Y., and Wong, H. (1995), "A Constitutive Model for Estimating Multiaxial Notch Strains",

ASME Journal of Engineering Materials and Technology, Vol. 117, No. 1, Jan., pp. 33-40.

McClintock, F. A. (1987), "Stastical Estimation: Linear Regression and the Single Variable", Research Memo 274, Fatigue and Plasticity Laboratory, Cambridge, Massachusettes Institute of Technology, Feb.

Neuber, H. (1961), "Theory of Stress Concentraion for Shear-Strained Prismatic Bodies with Arbitrary Nonlinear Stress-Strain Law", *ASME Journal of Applied Mechanics*, Vol. 28, No. 4, pp. 544-551.

Peterson, R. E. (1974), *Stress Concentration Factors*, John Wiley and Sons.

Ramberg, W., and Osgood, W. R. (1943), "Description of Stress-Strain Curves by Three Parameters", National Advisory Committee of Aeronautics, Technical Note 902.

Sharpe, W., Yang, C., and Tregoning, R. (1992), "An Evaluation of the Neuber and Glinka Relations for Monotonic Loading", *ASME Journal of Applied Mechanics*, Vol. 59, Oct., pp. S50-S56.

Table 1. Material Properties of 7075-T6 Aluminum Alloy

	Experimental	ASM Handbook [1985]	Percent Diff.
Yield Strength	505 MPa (73.3 ksi)	503 MPa (73.0 ksi)	0.4
Ultimate Strength	576 MPa (83.6 ksi)	572 MPa (83.0 ksi)	0.7
Elastic Modulus	69.6 GPa (10.1 msi)	71.0 GPa (10.3 msi)	2.0
Poisson's Ratio	0.37	0.33	12.1

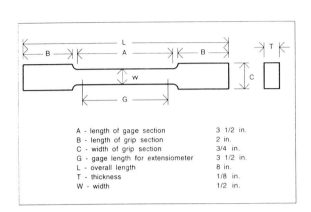

Figure 1. Tension Test Coupon

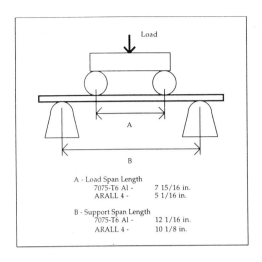

Figure 3. Four-Point Bending Test Apparatus

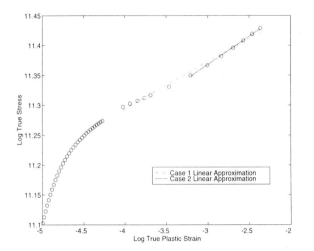

Figure 2. Log Plot of True Stress-Plastic Strain of Aluminim Alloy

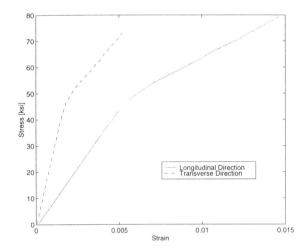

Figure 4. Stress-Strain Curves for ARALL-4

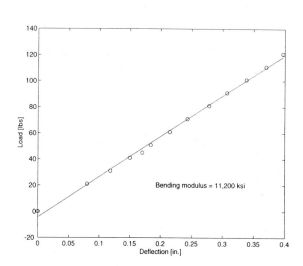

Figure 5. Load-Deflection Curve for Flexural Test of ARALL-4

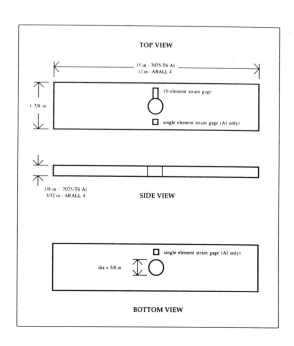

Figure 7. Circular Notched Sample Dimension and Strain Gage Locations

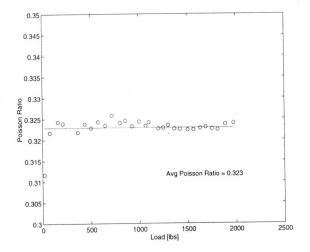

Figure 6. Poisson's Ratio of ARALL-4 with Fibers in Loading Direction

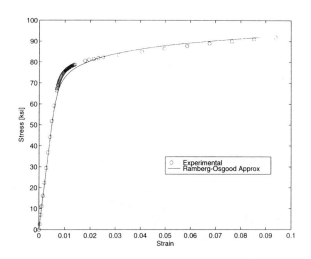

Figure 8. Stres-Strain Curve for Aluminum Alloy

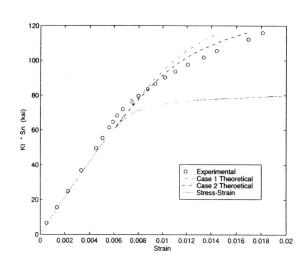

Figure 9. Comparison of Theoretical and Experimental Notch Strains in Aluminum Alloy for Two Different Cases of K and n Values

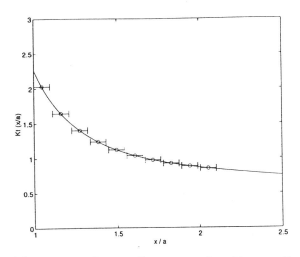

Figure 11. Stress Concentration Factor Gradient Away From the Hole for Aluminum Alloy. Individual Strain Gage Locations on the 10-Element Strip are Indicated by Horizontal Error Bars

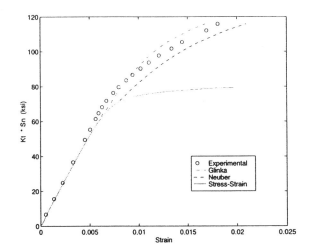

Figure 10. Compasrions of Glinka and Neuber Predictions for Notch Strains in Aluminum Alloy Under Tension

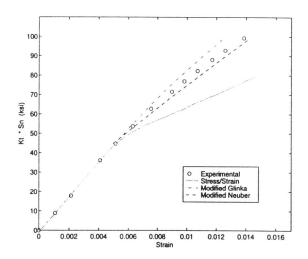

Figure 12. Comparison of Bilinear Glinka and Neuber Predictions for ARALL-4 Under Tension

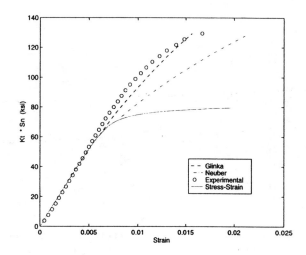

Figure 13. Comparison of Bilinear Glinka and Neuber Predictions for Aluminum Alloy Under Out-of-Plane Bending

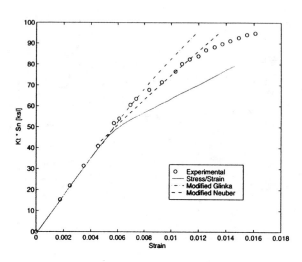

Figure 15. Comparison of Bilinear Glinka and Neuber Predictions for ARALL-4 Under Out-of-Plane Bending

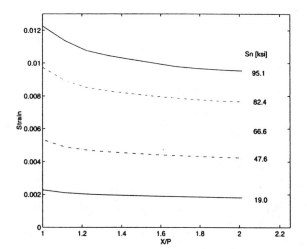

Figure 14. Strain Gradient Ahead of the Circular Notch for Several Nominal Stresses for Aluminum Alloy Under Out-of-Plane Bending

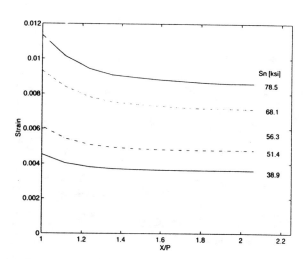

Figure 16. Strain Gradient Ahead of the Circular Notch for Several Nominal Stresses for ARALL-4 Under Out-of-Plane Bending

PVP-Vol. 356, Integrity of Structures and Fluid Systems,
Piping and Pipe Supports, and Pumps and Valves
ASME 1997

EXTREME VALUE PROBABILISTIC MODEL FOR FATIGUE LIFE IN BRITTLE PLATES WITH RANDOM CRACKS

David W. Nicholson, Ph.D.
Professor and Director

Peizhong Ni, Ph.D.

Institute for Computational Engineering
Department of Mechanical, Materials and Aerospace Engineering
University of Central Florida
Orlando, FL. 32816

ABSTRACT:

Fatigue life prediction is a subject of great interest for gas turbine components (Shih (1980), Tryon (1995)). Few if any models are available which rigorously apply fatigue criteria together with extreme value methods to predict fatigue life. The present study introduces a probabilistic model in which established criteria based on *fracture mechanics* are used for mixed-mode fatigue crack growth and eventual failure. The crack number, length and orientation at the onset of fatigue are random variables. Recently, for static fracture under *fixed* load, the authors introduced a conditional probability model leading to a probability density function (PDF) for the strength of brittle plates (more generally two dimensional media) with random cracks under mixed mode and multiaxial loading. (Ni (1996), Nicholson and Ni (1997-a), Nicholson and Ni (1997-b)). Unlike the few previous models involving extreme value methods (Trustrum (1977), Ichikawa (1995), She (1993)), the formulation is not restricted with respect to crack number or length. The present study extends the probabilistic model to fatigue life under fluctuating load. An *explicit* probability density function is derived for N_f, the number of cycles to *plate failure*. Biaxial loading and torsion are accommodated. The mean and variance of N_f are computed to show dependence on plate size, on the variance of the crack length distribution, on the multiaxial load factors, and on the parameters of the fatigue model.

1. INTRODUCTION

Fatigue life prediction is a subject of great interest for gas turbine components. Shih (1980) and Tryon (1995) consider application of probabilistic fatigue crack models to gas turbine engine components, but in general terms. High strength steels and ceramics are used increasingly in gas turbine components to achieve enhanced durability under severe operational stresses and temperatures. However, these materials are often brittle, so that their fracture and fatigue behavior is of great research interest.

Given the statistical nature of fatigue crack growth in brittle materials, a number of probabilistic models have been proposed (Wirshung (1995), Harris (1995), Bloom (1983)). Few if any models are available which rigorously apply fatigue and fracture models *from fracture mechanics* to cracks which have random number, length and orientation at the onset of fatigue. Perhaps the best known model is the Birnbaum-Saunders distribution (cf. Mann (1974), Castillo (1988)). It regards the incremental crack extension under a load as a random variable, with failure occurring when the crack attains a critical length; it does not use an established crack growth model from fracture mechanics, and takes no account of the random number, length or orientation of pre-existing cracks. Issues of fracture mode and multiaxial loading are not addressed. A number of more recent but similar models are discussed in Castillo (1988).

In contrast, the present study introduces a probabilistic model based on established criteria from *fracture mechanics*. The crack number, length and orientation at the onset of fatigue are treated as random variables. A model due to Sih and Barthelemy (1980), hereafter called the SB model, is used, with some further development, for mixed-mode fatigue crack growth under multiaxial loading, and eventual failure. Biaxial loading and torsion are accommodated.

Recently, for *static fracture* under *fixed* load, the authors introduced a conditional probability model based on a cell concept and the Sih (1974) mixed-mode static fracture criterion. A probability density function (PDF) was derived for the strength of brittle plates with random cracks (Ni (1996), Nicholson and Ni (1997-a), Nicholson and Ni (1997-b)). The strength distribution depended on plate size, the parameters of the crack length distribution, the parameters of the fracture criterion and the load ratios. Unlike related earlier models (Trustrum (1977), Ichikawa (1995), She (1993)), the model is applicable without restriction on crack number or length.

The present study extends the probabilistic model to fatigue life under fluctuating multiaxial load. At the onset of fatigue, the crack number follows a binomial distribution, crack length follows a *normalized* Gamma distribution, and crack orientation follows a uniform distribution. The SB model for mixed-mode fatigue crack growth is given additional development and is extended to biaxial load and torsion using the solution of Eftis and Subramonian (1978). Fracture occurs when at least one crack elongates (including kinking) and rotates to the point of satisfying the mixed-mode *static fracture* criterion, based on Sih (1974), for multiaxial loads. The use of a cell notion together with conditional probability relations allows the current model to avoid limitation to large numbers of long cracks. An extreme value probabilistic model is formulated, leading to an *explicit* probability density function for N_f, the number of cycles to *plate failure*. The mean and variance of N_f are computed to show dependence on plate size, on the variance of the crack length distribution, on the load ratios, and on the parameters of the fatigue model. Scaling relations are derived and computed.

2. MIXED MODE FATIGUE MODEL UNDER MULTIAXIAL LOADS

The following relations represent further development of the SB model for mixed mode fatigue crack growth in large plates. Crack growth is controlled by the fluctuation of the strain energy intensity factor S, initially expressed for uniaxial loading in Sih (1974). Here S is extended to biaxial load using the Eftis-Subramonian (1978) solution, and also accommodates torsion.

The model is based on the equivalent straight line crack from the SB model, as follows. Referring to Fig. 1, let a denote the instantaneous half-distance between the crack tips, with corresponding orientation β clockwise from the principal axis corresponding to the major in-plane principal stress. In the next increment of growth the (physical) crack will extend along the direction θ_0 from the crack axis. This gives rise to a' and β' for the equivalent straight crack. The following relations have been derived in this study.

$$da' = da \cos \theta_o \qquad (1)$$

$$d\beta = \tan \theta_o \frac{da'}{a'} . \qquad (2)$$

By way of proof, the equivalent crack after incremental elongation and rotation has length $a+da'$. From elementary geometry, using $a' = a$,

$$a + da' = a \cos d\beta + da \cos(\theta_o - d\beta) \qquad (3)$$

furnishing Eq.(1) to first order in dβ. Also,

$$a' \sin d\beta = da \sin\theta$$
$$= da' \tan\theta_0, \qquad (4)$$

furnishing Eq. (2). Hereafter the accent marks will not be displayed.

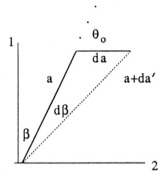

Fig.1. Equivalent Straight Line Crack

The S-B model assumes

$$\frac{da}{dN} = \Omega(\Delta S) \qquad (5)$$

where Ω is determined experimentally, and N is the number of cycles. A commonly assumed form is

$$\Omega(\Delta S) = \chi_{sb}(\Delta S)^{n_{sb}} \qquad (6)$$

where χ_{sb} and n_{sb} are constants determined experimentally. In the following we assume that S fluctuates between S and 0, and hence $\Delta S = S$, and the SB model is rewritten as

$$\frac{da}{dN} = \Omega(S) \ . \tag{7}$$

The strain energy intensity factor S suited for biaxial loading and torsion is derived in detail in Ni (1996) and Nicholson and Ni (1977-b). The coordinates r and θ are defined in Fig. 2 relative to the tip of the equivalent straight crack. S satisfies a relation of the form

$$S = \sigma^2 a \Psi(\beta, k_b, k_T, \theta_o(\beta, k_b, k_T)) \ . \tag{8}$$

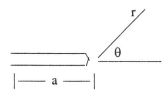

Fig. 2. Crack Tip Coordinate System

In particular, if K_I, K_{II} and K_{III} are the stress intensity factors in Mode I, II and III, respectively, then (Sih (1974))

$$S = a_{11}K_I^2 + 2a_{12}K_I K_{II} + a_{22}K_{II}^2 + a_{33}K_{III}^2 \tag{9}$$

$$a_{11} = [3-4\nu-\cos\theta](1+\cos(\theta))/16\pi\mu$$

$$a_{12} = [2\sin\theta(\cos\theta-(1-2\nu)]/16\pi\mu$$

$$a_{22} = [4(1-\nu)(1-\cos\theta)+(1+\cos\theta)(3\cos\theta-1)]$$

$$a_{33} = 4/16\pi\mu \ . \tag{10}$$

For mixed-mode fatigue crack growth, the direction of crack propagation from the axis of the equivalent crack coincides with the angle θ_o which minimizes the strain

energy intensity factor. In general, θ_o must be evaluated numerically as a function of β, k_b and k_T. The growth and rotation of the equivalent crack leads to catastrophic crack propagation when the static fracture criterion is satisfied, i.e., when $S = S_{cr}$ where S_{cr} is a material property called the critical strain energy intensity. For biaxial load and torsion, S is obtained as follows (Ni (1996), Nicholson and Ni (1997-b). Let

$$K_I = \sigma\sqrt{a}\kappa_I \qquad K_{II} = \sigma\sqrt{a}\kappa_{II} \qquad K_{III} = \sigma\sqrt{a}\kappa_{III}, \tag{11}$$

$$\Psi = \hat{a}_{11}\kappa_I^2 + 2\hat{a}_{12}\kappa_I\kappa_{II} + \hat{a}_{22}\kappa_{II}^2 + \hat{a}_{33}\kappa_{III}^2 \tag{12}$$

where

$$\kappa_I = \sqrt{\pi}[(1+k_b)-(1-k_b)\cos2\beta]/2 \qquad \kappa_{II} = \sqrt{\pi}(1-k_b)\sin2\beta/2$$

$$\kappa_{III} = \sqrt{\pi}k_T\sin\beta \tag{13}$$

and the circumflex ∧ implies evaluation at θ_o. The function Ψ above now depends on k_T as well as k_b. It follows that S has the form

$$S = a\sigma^2\Psi \ . \tag{14}$$

We note that the S-B model Eq.(7) is equivalent to

$$\frac{dS}{dN} = Y(S) \ . \tag{15}$$

if the function Y is chosen accordingly. In particular,

$$\frac{dS}{dN} = \Xi(S)\ \Omega(S) \qquad \Xi = \sigma^2[\Psi + \frac{d\Psi}{d\beta}\tan\theta_0] \ . \tag{16}$$

Now Ξ is a function of β. But, by integating Eqs. (2,5) β is obtained, at least numerically, as a function of S.

For the sake of a numerical example, we select the form

$$\frac{dS}{dN} = \chi S^n \ . \tag{17}$$

In the following we will consider the case in which n > 1. For suitable choices of χ and n, this model implies crack growth characteristics very similar to Eq. (7).

3. PROBABILISTIC MODEL

As illustrated in Fig. 3, we consider the plate to consist of N cells each of which has a probability of M/N of containing a center crack. Hence the crack number satisfies a binomial distribution in which M is the expected number of cracks. The crack initial length follows a normalized two parameter probability distribution, and we have selected a Gamma distribution for this purpose. Restrictions on admissible crack length distributions are derived in Ni (1996) and Nicholson and Ni (1997-a). The crack length PDF is denoted by $f(a_0,\alpha_1,\alpha_2)$ where a_0 is the initial crack length, and α_1 and α_2 are the mean and variance of the initial crack length. The corresponding cumulative distribution function (CDF) is denoted by $F(a_0,\alpha_1,\alpha_2)$. Finally the initial crack orientation angle β_0 follows a uniform distribution on $[0,\pi/2]$.

For purposes of a probabilistic model for failure ensuing from fatigue crack growth, we require the number of cycles to failure N_{fi} for the crack in the ith cell, and thereafter $Pr(N_{f,min} < N_f)$ where $N_{fmin} = \min_i N_{fi}$. For the crack in the ith cell with initial half-length a_{oi} and initial orientation β_{oi}, define $\Psi_{oi} = \Psi(\beta_{oi},k_b,k_T)$. Integration of Eq. (5) furnishes

$$\frac{\sigma^2\alpha_1}{S_{cr}}\frac{a_{0i}}{\alpha_1} = [\ 1 - \frac{\chi(1-n)}{S_{cr}^{1-n}}\ N_{fi}]^{\frac{1}{1-n}}/\Psi_{oi}\ . \tag{18}$$

The probability G, to be derived, that the plate will fail within N_f cycles equals the probability that $N_{fmin} \leq N_f$. Thus G also equals the probability that $a_{omax} \geq a_f$ for $0 < \beta_0 < \pi/2$, where $a_{omax} = \max a_{oi}$. The probability of survival of the crack in the ith cell after N_f cycles is

$$\Phi = \int_0^{\pi/2} \frac{2}{\pi}[1-F(\zeta,\alpha_1,\alpha_2,k_b,k_T)]d\beta \tag{19}$$

where $\zeta = \sigma^2\alpha_1 c/S_{cr}$ and $c = a_o/\alpha_1$. Also, for purposes of integration we have introduced the change of variables

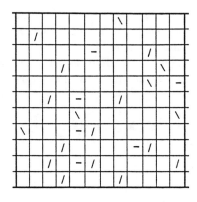

Fig 3. Plate with Center Cracked Cells

$$N_f = \frac{S_{cr}^{1-n}}{\chi(1-n)}\ [1-\zeta^{1-n}]\ . \tag{20}$$

Now Φ represents the probability of failure of one crack after N_f cycles, and $M\Phi/N$ represents the probability of failure of one cell. The probability of survival of all N cells, assuming $M/N \gg 1$ (Ni (1996), Nicholson and Ni (1997-a)) is

$$\overline{G} = 1-(1- M\Phi/N)^N \approx 1 -\exp(-M\Phi)\ . \tag{21}$$

However, the probability that none of the cells contains a crack is exp(-M). We now use the relations of conditional probability theory to conclude the following. *Provided that at least one cell contains a crack*, the probability of failure for the plate after N_f cycles is

$$G = \frac{1-\exp(-M\Phi)}{1-\exp(-M)}. \tag{22}$$

Note that $N_f \to 0$ as $\zeta \to \infty$, while $N_f \to 0$ as $\zeta \to \infty$. Also note that Φ is a function of ζ, k_b and k_T and of the parameters α_1 and α_2 of the crack length PDF. As discussed in Ni (1996), the PDF f representing a two parameter Gamma distribution can be normalized using the variate $c = a_0/\alpha_1$. The variance of a_0 is denoted by α_2, and the corresponding variance of c is denoted as $\Sigma_c = \alpha_2/\alpha_1^2 - 1$. To reflect this modification, Φ is replaced by Φ_c and G is replaced by G_c. Finally, the PDF corresponding to G, called the *extreme value distribution* (EVD) is given by

$$g_c = \frac{dG_c}{dN_f} = \frac{dG_c}{d\zeta}\frac{d\zeta}{dN_f} \ . \tag{23}$$

4. SCALING RELATIONS

Provided that f meets some restrictions discussed in Ni (1996), the mean $E(N_f)$ and variance $VAR(N_f)$ of the number of cycles to failure for the whole plate are given by

$$E(N_f) = \int_0^\infty (1-G_c)dN_f \qquad E(N_f^2) = 2\int_0^\infty N_f(1-G_c)dN_f$$

$$VAR(N_f) = E(N_f^2) - E^2(N_f) \ . \tag{24}$$

Of course they depend on M, Σ_c, k_b and k_T. Use of Eqs. (20,22) furnishes

$$E(N_f) = \frac{S_{cr}^{1-n}}{\chi} \int_0^\infty [1-G_c(\zeta)]\zeta^{-n}d\zeta \tag{25}$$

$$E(N_f^2) = \frac{2}{1-n}[\frac{S_{cr}^{1-n}}{\chi}]^2 \int_0^\infty [1-G_c(\zeta)][1-\zeta^{1-n}]\zeta^{-n}d\zeta \ .$$

It is of great practical interest to compute scaling relations between service and laboratory structures. Consider two plate populations with areas $A_1 > A_2$ and $M_1 > M_2$, and assume that $A_1/A_2 = M_1/M_2$. The subscripts 1 and 2 denote the service and laboratory plates, respectively. It is assumed that both experience the same value of σ, k_b and k_T, and contain pre-existing cracks coming from the same population. The ratios of the means and variances of the number of cycles of failure are to be calculated as a function of M, i.e.:

$$\frac{E_1}{E_2} = \frac{E(M_1,\Sigma_c,k_b,k_T)}{E(M_2,\Sigma_c,k_b,k_T)} \qquad \frac{VAR_1}{VAR_2} = \frac{VAR(M_1,\Sigma_c,k_b,k_T)}{VAR(M_2,\Sigma_c,k_b,k_T)} \ . \tag{26}$$

5. NUMERICAL RESULTS

A computer program has been developed the calculate the foregoing ratios. For AISI 4130 steel, Sih and Macdonald (1974) gives the following values: the elastic modulus is $E = 30\times10^6$ psi, Poisson's ratio is $v = 0.25$ and $S_{cr} = 32.2$ lb/in. Also let $\sigma = 170$ ksi. The values $\chi = 0.01$ and $n = 2$ have been chosen for the sake of illustration.

The foregoing expressions involve triple integration. They have been implemented using 20 node Gaussian quadrature for the finite intervals and 41 node Kronrod quadrature for the infinite intervals (Squire (1970)). The values chosen represent crack length variances $\Sigma_c = 5$, 4 and 3.

Figure 4 presents numerical results for the mean N_f (cycles to failure) ratio as a function of k_b and M_1/M_2, and Fig. 5 gives the corresponding results for the variance ratio. The values of Σ_c and k_T are 1 and $1/\sqrt{3}$. Figure 4 presents the corresponding results for VAR_1/VAR_2. In general the computations show expected trends. For example, the greater the variance in the crack length, the greater the decrease in the mean N_f ratio. Computations of the scaling relations are presented more extensively in Ni (1996).

5. CONCLUSION AND DISCUSSION

Previous probabilistic models for static fracture in brittle plates have been extended to predict the number-of-cycles to failure by fatigue crack growth. The Sih-Barthelemy model (1980) for mixed mode fatigue has been further developed and extended to multiaxial loads. The EVD of interest has been derived. Scaling relations have been formulated and computed. The model is believed to be applicable to gas turbine engine components.

6. REFERENCES

Bloom, J.M, and Ekvall, J.C., 1983, *Probabilistic Fracture Mechanics and Fatigue Methods: Application for Structural Design and Maintenance*, ASTM STP 798, Baltimore, MD.

Castillo, E., 1988, *Extreme Value Theory in Engineering*, Academic Press, San Diego, CA.

Eftis, J. and Subramonian, N., 1978, "The Inclined Crack under Biaxial Load", *Engineering Fracture Mechanics*, Vol 10, pp. 43-67.

Harris, D.O., 1995 ,"Probabilistic Fracture Mechanics", *Probabilistic Structural Mechanics Handbook*, ed. C. Sundararajan, Chapman and Hall Publ, New York.

Ichikawa, M., 1995, "Probability Evaluation of Brittle Solids Under Polyaxial Stress States Based on Maximum Energy Release Rate Criterion", *Engineering Fracture Mechanics*, VOl 51, No. 4, p. 629- 635.

Mann, N.R., Schafer, R.E. and Singpurwalla, N.D.,1974, *Methods for Statistical Analysis of Reliability and Life Data*, John Wiley and Sons, New York.

Ni, P., "Probabilistic Theory for Mixed-Mode Brittle Fracture and Fatigue with Random Cracks", Doctoral Dissertation, University of Central Florida, Orlando, FL 1996.

Nicholson, D.W. and Ni, P.,1997-a, "Extreme Value Probabilistic Theory for Mixed Mode Brittle Fracture", under review.

Nicholson, D.W. and Ni, P., 1997-b, "Extreme Value Probabilistic Model for Brittle Plates under Multiaxial Loads", this conference.

She, S. and Landes, J.D., "Statistical Analysis of Fracture in Graphite", *Int J. Fracture*, vol 63, p. 189-200, 1993.
Shih, T., "An Evaluation of the Probabilistic Approach to Brittle Design", *Engineering Fracture Mechanics*, Vol 13, pp. 257-271.

Sih, G., 1974, "Strain Energy Density Factor Applied to Mixed-Mode Crack Problems", *International Journal of Fracture*, Vol 10, No. 3, pp. 305-321.

Sih, G. and Barthelemy, B.M., 1980, "Mixed Mode Fatigue Crack Growth Predictions", *Engineering Fracture Mechanics*, Vol 13, pp. 439-451.

Sih, G. and Macdonald, B., 1974, "Fracture Mechanics Applied to Engineering Problems-Strain Energy Density Fracture Criterion", *Engineering Fracture Mechanics*, Vol 13, pp. 439-451.

Squire, W.B., 1970, *Integration for Engineers and Scientists*, American Elsevier, New York.

Tryon, R.G. et al 1995, "Development of a Reliability Based Fatigue Model for Gas Turbine Engine Structures", *Engineering Fracture Mechanics* Vol 53, No. 5, pp. 807-828.

Trustrum, K., and Jayatilaka, A. De S., 1977, "Application of Statistical Method to Brittle Fracture in Biaxial Loading Systems", *Journal of Materials Science*, Vol 12, pp. 2043-2048.

Wirshung, P. H., 1995, "Probabilistic Fatigue Analysis", *Probabilistic Structural Mechanics Handbook*, ed. C. Sundararajam, Chapman and Hall Publ, New York.

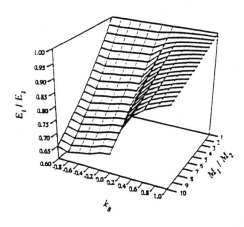

Fig. 4.   Mean N_f ratio vs M_1/M_2

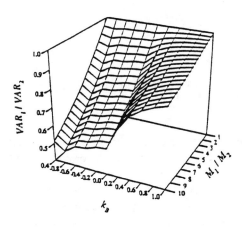

Fig. 5. Ratio of Variances of N_f vs. M_1/M_2

PVP-Vol. 356, Integrity of Structures and Fluid Systems,
Piping and Pipe Supports, and Pumps and Valves
ASME 1997

EXTREME VALUE PROBABILISTIC MODEL FOR BRITTLE PLATES WITH RANDOM CRACKS UNDER MULTIAXIAL LOADS

David W. Nicholson, Ph.D.
Professor and Director

Peizhong Ni, Ph.D.

Institute for Computational Engineering
Department of Mechanical, Materials and Aerospace Engineering
University of Central Florida
Orlando, FL. 32816

ABSTRACT

Several authors (Shih (1980), Tryon (1995)) have discussed the application of probabilistic fracture mechanics (PFM) to gas turbine structures, but in general terms. The literature contains many PFM models, most of which assume that plates contain cracks with *random strengths*. Catastrophic fracture occurs at the weakest crack, implying a probability distribution for the strength of the plates. Criteria from fracture mechanics, which take account of *crack number, length and orientation*, are not applied explicitly in such models. In contrast, a few models have assumed the presence of cracks with random lengths and orientations, and have explicitly applied fracture criteria. The ensuing strength distributions for the plates depend on the parameters of the crack length distribution, on the fracture criteria, and on the plate size. However, the models have several limitations, including: (1) a probabilistic formulation which depends on large numbers of long cracks being present in the plates; (2) uniaxial loading; and (3) simplified fracture criteria. Recently, under the restriction of uniaxial loading, the authors have introduced a conditional probability model based on a cell concept, leading to a distribution which is applicable without restrictions on crack number or length. In the present study, the formulation is extended to the general case of biaxial loading and torsion (Mode III). Expressions for the mean and variance of the strength are derived in terms of plate size, the crack length distribution, the fracture criteria and ratios of the applied loads. Scaling relations are computed and presented.

1. INTRODUCTION

Several authors (Shih (1980), Tryon (1995)) have discussed the application of probabilistic fracture mechanics (PFM) to gas turbine structures, but in general terms. The overall goal of the current research is to formulate a PFM model which is suited for components, for example of gas turbine systems, which have a random distribution of flaws and for which there is a need to assess the probability of failure. A large number of PFM models have been reported in the literature (Harris (1995), Bloom (1983), She (1993)). Many directly assume that the cracks have random strengths. Only a few address the distribution of crack lengths and orientations and rigorously apply fracture mechanics together with extreme value probability theory (weakest link theory). Two exceptions are Trustrum (1977), Ichikawa (1995), which address asymptotic behavior by assuming large numbers of long cracks. In Nicholson and Ni (1997) and Ni (1996), under the restriction to uniaxial loading, the cell concept and conditional probability relations have been used to develop a formulation which applies without restriction on crack number or length. In the present study the formulation is extended to the general case of biaxial loading plus torsion.

The model is based on the weakest link notion and is thus limited to brittle materials. Some advanced steels and modern ceramics offer high strength and temperature resistance. Accordingly, they have been applied to components of gas turbine engines. However such materials can have an inherent disadvantage - brittleness. Probabilistic models are a powerful tool for this purpose and are investigated here. They may be viewed as having three fundamental elements: (a) appropriate probability density functions (in the current investigation for crack number, length and orientation); (b) fracture criteria (in the current investigation for mixed mode fracture under multiaxial load); and (c) the ensuing extreme value for the strength distribution of the structure.

2. EXTREME VALUE DISTRIBUTION FOR BRITTLE FRACTURE

2.1 Single Crack

Referring to Fig. 1, consider a plate (more generally a two-dimensional body) of brittle material under biaxial loads and torsion, consisting of center-cracked cells. The cracks come from a population whose lengths follow a two parameter probability density function $f(a; \alpha_1, \alpha_2)$, where a is the crack half-length and α_1, α_2 are distribution parameters. Clearly f is one-sided, since $a \geq 0$. We also assume that $\alpha_1 \geq 0$ and $\alpha_2 \geq 0$. The cumulative distribution function (CDF) corresponding to f is denoted as $F(a; \alpha_1, \alpha_2)$. We interpret α_1 and α_2 as the mean and variance of a:

$$\alpha_1 = \bar{a} = E(a) \quad \alpha_2 = E((a-\alpha_1)^2) \quad E(a) = \int_0^\infty a\,da \quad (1)$$

and $E(.)$ denotes the expectation. In the following sections, for notational convenience, the parameters α_1 and α_2 will not be displayed in f and F.

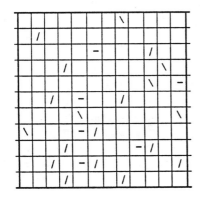

Fig. 1. Plate with N cells and M inclined center-cracks

We now suppose that the plate consists of N cells of equal area, each of which has a probability M/N of containing a crack, where M < N. Only one crack is permitted in a cell, and the cracks may not cross cell boundaries prior to fracture. It follows that the number of cracks in N cells satisfies a binomial distribution, in which M is the expected number of cracks.

We momentarily digress to consider the burden imposed by the restriction to large N/M. Clearly, the cell must be much larger in some sense than the critical crack. For illustration, suppose M/N =1/100 and M = 10. Then $(1-M/N)^N$ = 4.31e-5 while exp(M) = 4.54e-5, for a 5% error. Now suppose the plate has dimensions L by L and the minimum cell size is set at $25\,a_f^2$, assuming square cells. Then $N_{max} = (L/a_f)^2/25$. Further if N_{max} = 100 M, then the asymptotic conditions are attained if the restriction $L > 50 \sqrt{M}\,a_f$ is satisfied.

Now let β denote the angle of the crack from the tensile axis, as shown in Fig 1. We assume that a and β are statistically independent and that all crack orientations are equally likely between $\beta = 0$ and $\beta = \pi/2$. (Actually, M cracks between $-\pi/2$ and $\pi/2$ are equivalent, from a fracture mechanics viewpoint, to M cracks between 0 and $\pi/2$). Hence the crack orientation follows a uniform distribution $f_\beta = 2/\pi$. Accordingly the distribution of random cracks can be expressed by the *joint probability density function*

$$f_{a\beta}(a,\beta) = \frac{2}{\pi} f(a) \quad (2)$$

where for simplicity of notation the distribution parameters α_1 and α_2 are no longer displayed.

The probability of failure for *one crack* at stress σ is

$$\Phi(\sigma) = \Pr(\sigma_f \leq \sigma)$$

$$= \Pr(a \geq a_f(\sigma,\beta))$$

$$= 1 - \Pr(a < a_f(\sigma,\beta))$$

$$= 1 - \frac{2}{\pi}\int_0^{\pi/2} F(a_f(\sigma,\beta))d\beta \quad F(a_f) = \int_0^{a_f} f(a)da \quad (3)$$

where a_f is obtained from the fracture criterion as a function of σ and β. In particular, the Sih (1974) criterion will be shown to imply the form

$$a_f = \frac{S_{cr}}{\sigma^2 \Psi(\beta, k_b, k_T)} \quad (4)$$

where, in this presentation, S_{cr} is a material constant called the critical strain energy intensity factor. The function $\Psi(\beta, k_b, k_T)$ and the parameters k_b and k_T express the fracture criterion and are defined in a subsequent section. Note that $\sigma \to 0$ corresponds to $a_f \to \infty$, and $\sigma \to \infty$ to $a_f \to 0$.

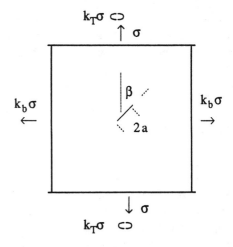

Fig. 2. Inclined crack under biaxial loads and torsion

2.2 Extreme Value Distribution: Multiple Cracks

The function Φ in Eq.(3) applies to one crack under stress σ. We now consider the strength of the whole plate, which contains N cells each of which has a probability M/N of containing a crack. We assume that N/M >> 1. As previously stated, M is the expected number of cracks in the plate. The probability that a cell does not contain a crack is 1 - M/N; the probability that none of the cells contains a crack is

$$(1 - M/N)^N \cong \exp(-M) \qquad (5)$$

and hence 1-exp(-M) is the probability that at least one cell contains a crack.

The probability that the ith cell will fail under stress σ is $M\Phi(\sigma)/N$, and the probability it will survive is $1 - M\Phi(\sigma)/N$. The probability that all cells in the plate will survive is

$$(1-M\Phi(\sigma)/N)^N \cong \exp(-M\Phi(\sigma)), \qquad (6)$$

and the probability that at least one cell will fail is $\overline{G}$ given by

$$\overline{G}(\sigma) = 1 - \exp(-M\Phi(\sigma)) . \qquad (7)$$

Note that $\sigma \to 0$ implies that $\Phi \to 0$ and $\overline{G} \to 0$. However, $\sigma \to \infty$ implies that $\Phi \to 1$ and $\overline{G} \to 1 - \exp(-M)$, which is the probability that at least one cell contains a crack. In this form $\overline{G}$ is not a CDF since it does not span the range $0 \le \overline{G} \le 1$, and is not suitable for computing the mean or variance of the strength. For this reason we introduce a *conditional probability distribution*. Now, *provided that at least one cell contains a crack*, the probability $G(\sigma)$ of plate failure under stress σ is

$$G(\sigma) = \frac{1-\exp(-M\Phi(\sigma))}{1-\exp(-M)} . \qquad (8)$$

The corresponding PDF, which we call the extreme value distribution (EVD) for the strength, is given by $g(s_f)$ where

$$g(s_f) = dG/ds_f \qquad G(\sigma) = \int_0^\sigma g(s_f)ds_f . \qquad (9)$$

Now we derive expressions for the mean and variance of the strength of the plate, as follows.

$$E(s_f) = \int_0^\infty s_f g\, ds_f$$

$$= \int_0^\infty (1-G)ds_f$$

$$= X(M;\alpha_1,\alpha_2) ; \qquad (10)$$

$$E(s_f^2) = \int_0^\infty s_f^2 g\, ds_f$$

$$= 2\int_0^\infty s_f(1-G)ds_f ;$$

$$VAR(s_f) = 2\int_0^\infty s_f(1-G)ds_f - \left[\int_0^\infty (1-G)ds_f\right]^2 \qquad (11)$$

$$= Y(M;\alpha_1,\alpha_2) , \qquad (12)$$

where X and Y have been introduced for later notational convenience. We have tacitly assumed that $s_f^2(1-G) \to 0$ as $s_f \to \infty$, which in fact is a condition under which the variance exists. This imposes a restriction on the types of functions that can be used as the crack length PDF (Ni, 1996).

2.3 Scaling Relations

In this section, we obtain scaling relations in the form of ratios of the mean and variance of the strength of a service structure to those of corresponding laboratory specimens. The goal is to estimate the strength of the service structure from the measured strength of the laboratory specimens. (We have not been able to locate conclusive experimental studies on scaling.) It is assumed that, in both structures, the cracks come from the same population, the cells are of equal area, and the probability that a cell contains a crack is M/N.

Consider two plate populations, "1" and "2", with areas $A_1 > A_2$ and expected crack numbers $M_1 > M_2$, and assume that

$$A_1/A_2 = M_1/M_2 . \qquad (13)$$

Let subscript 1 denote the service plates and subscript 2 the laboratory plates. Referring to Eqs.(10-12), we introduce the ratios

$$\frac{E_1(s_f)}{E_2(s_f)} = \frac{X_1(M_1;\alpha_1,\alpha_2)}{X_2(M_2;\alpha_1,\alpha_2)} \qquad \frac{VAR_1(s_f)}{VAR_2(s_f)} = \frac{Y_1(M_1;\alpha_1,\alpha_2)}{Y_2(M_2;\alpha_1,\alpha_2)} . \quad (14)$$

We observe that the foregoing ratios depend on the expected crack numbers M_1 and M_2, as well as the parameters α_1 and α_2 from the crack length distribution. We will shortly see that the mean crack length α_1 can be factored out of these ratios, thereby allowing the scaling ratios to be expressed purely in terms of a normalized crack length variance.

2.4 Normalized Crack Length Distribution

It is claimed in Jayatilaka (1979) that pre-existing cracks in brittle plates follow a Gamma distribution, which we also assume here. However, any distribution satisfying the restrictions derived in Ni (1996) using the cell concept is admissible. Other investigators have considered the logarithmic normal and the two-parameter Weibull distribution (Harris, 1995). Under additional restrictions on their parameters, these distributions are admissible in the current formulation.

The Gamma distribution with parameters $p > 0$ and $q > 0$ is expressed as

$$f(a) = \frac{q^p}{\Gamma(p)} a^{p-1} exp(-qa) \qquad \Gamma(p) = \int_0^\infty x^{p-1} exp(-x)dx . \quad (15)$$

The mean and variance of a are

$$\alpha_1 = p/q \qquad\qquad \alpha_2 = p/q^2 . \quad (16)$$

We introduce the normalized crack length $c = a/\alpha_1$. The mean of c is unity, and the variance of c is denoted as $\Sigma_c = \alpha_2/\alpha_1^2 -1$. Now denote the PDF of c as f_c, and elementary procedures imply that

$$f_c(c,\Sigma_c) = \alpha_1 f(a,\alpha_1,\alpha_2) . \quad (17)$$

The fracture criterion Eq. (4) is now rewritten as

$$c_f = a_f/\alpha_1 = \frac{s_{fc}^2}{\sigma^2\Psi} \qquad s_{fc} = \sqrt{S_{cr}/\alpha_1} . \quad (18)$$

Following the foregoing derivation of F, Φ, G and g, we obtain

$$F_c(\frac{s_{fc}^2}{\sigma^2\Psi(\beta)};\Sigma_c) = \int_0^{c_f} f(c;\Sigma_c)dc$$

$$\Phi_c(\sigma) = 1 - \frac{2}{\pi}\int_0^{\pi/2} F_c(\frac{s_{fc}^2}{\sigma^2\Psi(\beta)};\Sigma_c)d\beta$$

$$G_c(\sigma) = \frac{1-exp(-M\Phi_c(\sigma))}{1-exp(-M)}$$

$$g_c(s_f) = \frac{dG_c}{ds_f} . \quad (19)$$

Clearly G_c can be viewed as a function of the dimensionless variable $\zeta = s_f/s_{fc}$, so that Eqs. (10-12) can be rewritten as

$$E(s_f) = s_{fc}X_c(M,\Sigma_c)$$

$$X_c(M,\Sigma_c) = \int_0^\infty (1-G_c(\zeta;M,\Sigma_c))d\zeta$$

$$VAR(s_f) = s_{fc}^2 Y_c(M,\Sigma_c)$$

$$Y_c(M,\Sigma_c) = 2\int_0^\infty \zeta(1-G_c)d\zeta - \left[\int_0^\infty (1-G_c)d\zeta\right]^2 . \quad (20)$$

The scaling ratios now explicitly depend only on M_1, M_2 and Σ_c. In particular, referring to Eq. (14),

$$\frac{E_1(s_f)}{E_2(s_f)} = \frac{X_c(M_1,\Sigma_c)}{X_c(M_2,\Sigma_c)} \qquad \frac{VAR_1(s_f)}{VAR_2(s_f)} = \frac{Y_c(M_1,\Sigma_c)}{Y_c(M_2,\Sigma_c)} . \quad (21)$$

Finally, note that these expressions are valid for any admissible PDF.

3. FRACTURE CRITERION FOR BIAXIAL LOADS AND TORSION

For mixed mode a well established criterion is due to Sih (1974) in terms of a critical strain energy intensity factor. In the case of biaxial loading, the two-term solution due to Eftis et al (1978) gives the correct stresses. Accordingly, we extend Sih's criterion to biaxial load and torsion. The 1-axis corresponds to the direction of the maximum principal stress in the plane, and let $\sigma = \sigma_{11}$. The angle β is drawn clockwise from the 1-axis to the crack, as shown in Fig. 1. The stress σ_{22} is written as $k_b\sigma$, where k_b is called the biaxiality factor. Torsion (Mode III) corresponds to the stress $\sigma_{13} = k_T\sigma$, where k_T is called the torsional factor.

Suppose the strain energy stored in a differential volume element with unit thickness is given by $U(r,\theta)$. We define the strain energy intensity factor S using

$$S = \lim_{r \to 0} [r \frac{dU}{dA}] . \qquad (22)$$

It is necessary to know how dU/dA depends on r and θ in the vicinity of the crack tip. In particular,

$$dU/dA = A/r + B\sqrt{r} + C + O(\sqrt{r}) \qquad (23)$$

where A, B and C are independent of r and depend on the angle θ between the position vector and the crack axis. Only A contributes to S and is given by

$$S = A = a_{11}K_I^2 + 2a_{12}K_IK_{II} + a_{22}K_{II}^2 + a_{33}K_{III}^2 \qquad (24)$$

$$a_{11} = [3-4\nu-\cos\theta](1+\cos(\theta))/16\pi\mu$$

$$a_{12} = [2\sin\theta(\cos\theta-(1-2\nu))/16\pi\mu$$

$$a_{22} = [4(1-\nu)(1-\cos\theta)+(1+\cos\theta)(3\cos\theta-1)]$$

$$a_{33} = 4/16\pi\mu . \qquad (25)$$

For mixed-mode brittle fracture, the strain energy intensity criterion may be stated in terms of two basic hypotheses.

1. The direction of crack propagation from the crack axis coincides with the angle θ_o which minimizes the strain energy intensity factor.

2. Failure occurs when the strain energy intensity factor corresponding to $\theta_o(\beta)$ attains a critical value, denoted as S_{cr}

In general, θ_o must be evaluated numerically as a function of β, k_b and k_T. Let $K_I = \sigma\sqrt{a}\kappa_I$, $K_{II} = \sigma\sqrt{a}\kappa_{II}$, $K_{III} = \sigma\sqrt{a}\kappa_{III}$, and $\Psi = \hat{a}_{11}\kappa_I^2 + 2\hat{a}_{12}\kappa_I\kappa_{II} + \hat{a}_{22}\kappa_{II}^2 + \hat{a}_{33}\kappa_{III}^2$, where

$$\kappa_I = \sqrt{\pi}[(1+k_b)-(1-k_b)\cos2\beta]/2 \quad \kappa_{II} = \sqrt{\pi}(1-k_b)\sin2\beta/2 .$$

$$\kappa_{III} = \sqrt{\pi}k_T\sin\beta. \qquad (26)$$

and the circumflex $\wedge$ implies evaluation at θ_o. The function Ψ above now depends on k_T as well as k_b and is given by

$$\Psi = \hat{a}_{11}\kappa_I^2 + 2\hat{a}_{12}\kappa_I\kappa_{II} + \hat{a}_{22}\kappa_{II}^2 + \hat{a}_{33}\kappa_{III}^2 . \qquad (27)$$

4. Numerical Example and Discussion

A computer program has been developed to calculate the triple integrals involved in Eq. (19) in terms of the variance of the crack length distribution, k_b and k_T. For AISI 4130 steel the required parameters are obtained from Sih et al (1974): $E = 30 \times 10^6$ psi; $\nu = .25$ and $S_{cr} = 33.2$ lb-in. The numerical integrations have been implemented using Gaussian quadrature with 20 nodes for the finite intervals, and the associated Kronrod formula with 41 nodes for infinite intervals (Squire, 1970).

Owing to the use of c instead of a, the mean crack length α_1 has no effect on the scaling ratios. We take p from the normalized Gamma distribution to be 0.2, 0.5, 1.0, and 2.0, corresponding to the variances $\Sigma_c = 5.0$, 2.0, 1.0, and 0.5, respectively. Figure 3 displays the ratio of mean strength E_1/E_2 as a function of k_b and M_1/M_2, with $\Sigma_c = 1$ and $k_T = 1/\sqrt{3}$. Figure 4 is the corresponding plot for VAR_1/VAR_2. The trends displayed are as expected. For example, the mean strength ratio decreases strongly with increasing Σ_c. A much more extensive presentation of numerical results is given in Ni (1996).

5. CONCLUSION

A probabilistic model has been developed which combines rigorous fracture criteria with extreme value methods to describe the strength of brittle elastic plates containing pre-existing cracks with random number, length and orientation. Existing mixed-mode fracture criteria have been extended to accommodate biaxial loads plus torsion. Ratios representing statistical effects on scaling are formulated and computed. The model is believed to be applicable to gas turbine components, following the discussion in Shih (1980) and Tryon (1995).

6. REFERENCES

Bloom, J.M, and Ekvall, J.C., 1983, *Probabilistic Fracture Mechanics and Fatigue Methods: Application for Structural Design and Maintenance*, ASTM STP 798, Baltimore, MD.

Eftis, J. and Subramonian, N., 1978, "The Inclined Crack under Biaxial Load", *Engineering Fracture Mechanics*, Vol 10, pp. 43-67.

Harris, D.O., 1995 ,"Probabilistic Fracture Mechanics", *Probabilistic Structural Mechanics Handbook*, ed. C. Sundararajan, Chapman and Hall Publ, New York.

Ichikawa, M., 1995, "Probability Evaluation of Brittle Solids Under Polyaxial Stress States Based on Maximum Energy Release Rate Criterion", *Engineering Fracture Mechanics*, VOl 51, No. 4, p. 629- 635.

Jayatilaka, A. De S., 1979, *Fracture of Engineering Brittle Materials*, Applied Science Publishers LTD, London.

Nicholson, D.W. and Ni, P.,1997, "Extreme Value Probabilistic Theory for Mixed Mode Brittle Fracture", under review.

Ni, P., "Probabilistic Theory for Mixed-Mode Brittle Fracture and Fatigue with Random Cracks", Doctoral Dissertation, University of Central Florida, Orlando, FL 1996.

She, S. and Landes, J.D., "Statistical Analysis of Fracture in Graphite", *Int J. Fracture*, vol 63, p. 189-200, 1993.

Shih, T., "An Evaluation of the Probabilistic Approach to Brittle Design", *Engineering Fracture Mechanics,* Vol 13, pp. 257-271.

Sih, G., 1974, "Strain Energy Density Factor Applied to Mixed-Mode Crack Problems", *International Journal of Fracture*, Vol 10, No. 3, pp. 305-321.

Sih, G. and Macdonald, B., 1974, "Fracture Mechanics Applied to Engineering Problems-Strain Energy Density Fracture Criterion", *Engineering Fracture Mechanics*, Vol 13, pp. 439-451.

Squire, W.B., 1970, *Integration for Engineers and Scientists*, American Elsevier, New York.

Tryon, R.G. et al, 1995, "Development of a Reliability Based Fatigue Model for Gas Turbine Engine Structures", *Engineering Fracture Mechanics* Vol 53, No. 5, pp. 807-828.

Trustrum, K., and Jayatilaka, A. De S., 1977, "Application of Statistical Method to Brittle Fracture in Biaxial Loading Systems", *Journal of Materials Science*, Vol 12, pp. 2043-2048.

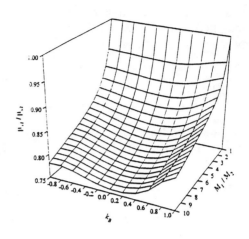

Fig. 3. Mean Strength vs M_1/M_2 and k_b
($k_T = 1/\sqrt{3}$; $\Sigma_c = 5.0, 2.0, 1.0, 0.5$)

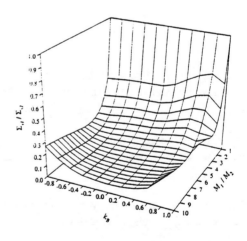

Fig. 4. Variance of Strength vs M_1/M_2 and k_b
($k_T = 1/\sqrt{3}$; $\Sigma_c = 5.0, 2.0, 1.0, 0.5$)

PVP-Vol. 356, Integrity of Structures and Fluid Systems,
Piping and Pipe Supports, and Pumps and Valves
ASME 1997

SAFE COMPRESSOR OPERATIONS: GENERAL APPROACH AND STANDARDS OVERVIEW

Boris Blyukher
Health and Safety Department
Indiana State University
Terre Haute, Indiana

ABSTRACT

It is well recognized that compressors operations involve significant risk to employees due to mechanical, electrical, high pressure, noise, and atmospheric hazards. One of the most common hazards is oxygen deficiency hazard (ODH) taking place due to undesired release and displacement of air with compressed gases. Information on compressor safety practices, however, has not been readily available to all who design, install, or operate compressor installations.

This paper will summarize the basic safety requirements of compressor systems commonly used in the process industries, R&D and test facilities. In developing these requirements the author has consulted many sources (OSHA , ANSI, and industry standards and regulations) and the resulting recommendations are believed to represent a summation of the best experience available. The paper will also generalize international experience gained by developing and establishing a compressor safety program at Superconducting Super Collider Laboratory (SSCL) and several cryogenic companies in Russia and Ukraine.

Specific recommendations cover the requirements for safety devices and protective facilities to prevent compressor accidents as a result of excessive pressure, destructive mechanical failures, internal fires or explosions, and leakage of toxic or flammable fluids. General safety practices and hazards unique to compressors are also covered. These principles promoting safe operation of compressors used for process industries are not restricted to new compressor systems.

INTRODUCTION

The use of compressors and compressed gases has a long history in the process industries. Experience has proven that the provision of certain basic safety features for compressor installations and operations are essential to minimize the possibility of destructive mechanical failures, electrical, fire, and ODH, or other serious accidents.

Industry practices that have evolved from this experience provide a sound basis for safety requirements of gas compression facilities.

The basic safety philosophy, the rules to be employed in the use of compressors, and the principal hazards associated with compressors considered here are outlined in OSHA Regulations 29 CFR, 1910. 101, *Compressed Gases (general requirements)*, 1910.219, *Mechanical Power-Transmission Apparatus*, 29 CFR, 1926, Subpart V, *Power Transmission and Distribution;* in ASME standards B15.1, *Safety Standard for Mechanical Power Transmission Apparatus*, B 19.3-1991 *"Safety Standards for Compressors for Process Industries"*, B 19.1-1990 *"Safety Standards for Air Compressors Systems"*, and horizontal industries standards API Standard 618, *"Reciprocating Compressors for General Refinery Services"* and others (see Bibliography).

In addition to standard technical requirements of safe operations, the author outlines administrative control and safety management procedures used for different industrial and R&D, and test facilities and applied to the compressor and its auxiliaries, including drivers, intercoolers, surge chambers, disengaging drums or scrubbers, interconnecting piping and lubrication, seal oil, and jacket water systems. For example, the Cryogenic Refrigerator Compressor Building at SSCL houses the Sullair Model C oil-flooded screw compressor packages that are used for the closed loop of Helium Refrigeration/Liquefaction Plants. The package includes open drip-proof motor, oil cooler and aftercooler, complete oil system including dual strainers and filters, reservoir and separator with coalescers, inlet and discharge check valves and inlet strainer, slide valve capacity control system, control panel containing pressure and temperature instrumentation, protective switches and capacity controls, closed transition starter with control power transformer and circuit breaker disconnects, oil removal system with gas management controls, all mounted on a structural steel base suitable for installation on a flat concrete pad.

The following potential hazards associated with operational procedures with compressors and proper mitigation measures are discussed: noise, heat burns, mechanical hazards associated with rotating equipment, remote possibility of oil or cryogen spill or leak, pressure, ODH, and possibility of fire.

COMPRESSOR SAFETY PROGRAM ELEMENTS

Compressor operations safety program elements include safety management at different levels of the organization, general requirements for safe operations and maintenance, "good" operating practices, and the startup, shutdown, and emergency procedures. Also included are methods of fire protection, safeguarding, electrical hazards and pressure release control, protection against the effects of noise exposure, compressor system and parts testing, environmental and personal monitoring, and control to increase safety in hazardous situations due to the presence of toxic or flammable gases, high pressures and/or high operating temperatures.

COMPRESSOR SAFETY MANAGEMENT

The total safety management effort for complex compressor systems involves several levels of organization (Fig. 1):

Level One. Corporate or headquarters. The safety management effort at this level usually consists of general oversight of multiple programs and development of policies and standards.

Level Two. Procurement activity. At the contracting level, the safety management task is to convert compressor safety requirements into contractual specifications.

Level Three. Safety engineering. At this level, management provides program oversight, policy, direction, and resource allocation.

Level Four. Contractor's engineering system safety program. This level is the working level for the system safety engineer and the program system safety effort.

Level Five. Specifications and requirements. This level tends to be compliance oriented. Designers and engineers incorporate safety codes, standards, and regulations. Shipping/Receiving Department inspects all incoming compressors and compressor auxiliary equipment. In the event that a compressor packages' unit is rejected, the intended user should be notified (Fig. 1c).

Level Six. Operational location. At this end user level, line supervisors and operators provide user input and review (Fig. 1b).

Line Supervisors are responsible for ensuring the direct implementation of safety procedures for all activities involving compressor systems operating under their control. Specifically they shall:

- ensure that the design of the compressor system (package) for installation within their jurisdiction, is submitted in writing to the Safety Department (Group);
- ensure that each worker is qualified (trained) on the compressor system (package) equipment in safety practices before the individual is permitted to perform tasks unsupervised;

- ensure that employees are aware of the requirements of safety procedure, are properly trained to operate equipment in a safe manner, and wear (use) adequate personnel protection equipment;
- ensure that adequate operating procedures and records are maintained.

Operators (individual workers) working with compressor(s) are responsible for following safe working procedures and for reporting to their supervisor any near misses, close calls, unusual event, or mishaps.

Safety specialists are responsible for assisting in the implementation of safety procedure (Fig. 1a). Specifically, they shall:

- conduct safety reviews;
- audit safety manuals, and test, check and operations programs pertaining to the compressor system.

GENERAL REQUIREMENTS FOR COMPRESSOR SAFE OPERATION AND MAINTENANCE

Operating Practices

Typical aspects of operating are considered with their correlation to safety and maintenance required by users to achieve their growing requirements for efficiency, reliability, environment, and safety. Good operational, maintenance and repair procedures can contribute to the safety of the operating crew as well as maintenance personnel. Therefore, supervisors should establish comprehensive maintenance and operating procedures with periodic reviews with affected personnel and instructions for new hires. These procedures should cover startup, break-in period, routine operation, routine maintenance, preventive maintenance, trouble shooting, and overhaul with the manufacturer's instructions as a basis.

Competent observation of compressor performance is one of the best methods of determining need for maintenance which, in turn, can be the best safety precaution available. Any damage observed or suspected should be reported to supervisors. If the condition impairs safe operation, the machine shall be taken out of service for repair in the prescribed safe manner. Safeguards that have been altered or damaged should be reported so appropriate action can be taken to ensure against worker injury.

The compressor package, including all of the components, shall be operated and maintained in accordance with operating instructions furnished by the manufacturers. Complete operating manuals for the compressor system shall include operating instructions for all components and machines purchased from other vendors. These instructions shall be retained and copies kept in a suitable place for easy reference.

Main fragments of general requirements for safe compressor operations and maintenance are presented and discussed in corresponding standards and regulations showed in Fig. 2.

- All operating personnel must become familiar with the physiological, physical, and chemical hazards associated with each compressed gas of interest. Each individual should work closely with experienced personnel until he becomes thoroughly familiar with all the equipment and procedures to be employed.

- Because of compressor facilities specific hazards (ODH, over-pressurization, fire, noise) access shall be strictly controlled. Visitors entering the area shall be escorted. The staff escorts shall be trained in safety requirements and familiar with the equipment, operation and hazards of the compressor facilities.
- A nondestructive examination and inspection program for highly stressed compressor parts shall be established and implemented on a predetermined schedule based on manufacturer's recommendation, which should be predicated on type, age, severity of service, and past industry performance and experience of equipment.
- External surfaces subject to temperatures in excess or 175°F (80°C) with which personnel may have contact shall be guarded or insulated.
- Rotating equipment shall not be placed in regular operation unless all safety features required by this Standard are in place.
- Color coding or other marking of piping systems is recommended. Piping system markings shall comply with OSHA Section 1910.144 and 1910.145. Coding to ANSI/ASME A13.1, ISO R508, or labels identifying line contents are preferred.
- Electric motors, their controls, and wiring shall be designed and installed in accordance with NFPA 70 and NEMA MG1, MG2, ICS-1, and ICS-2. Motors should be maintained in accordance with the manufacturer's instructions and NFPA 70B. Should a conflict exist, the manufacturer's instructions should be given preference. Safety requirements shall be in accordance with NFPA 70E.
- Compressors handling toxic or flammable gases shall be isolated from the process piping by means of blinds or double valves and bleeder when major maintenance is required. Before opening such compressors, the equipment should be purged or evacuated. Minor adjustments to the compressor and drive, such as rod packing and valve inspection (on an individual basis), may be performed without blinding provided adequate precautions, such as depressuring, are taken to protect personnel. Check valves shall not be relied upon for isolating compressors.

The following safety precautions are offered as a guide which, if conscientiously followed, will minimize the possibility of accidents throughout the useful life of this equipment.
- Installation, use and operation of the compressor shall be conducted only in full compliance with all pertinent OSHA regulations and all applicable Federal and industrial codes, standards and regulations (Fig. 2).
- The compressor should be operated only by those who have been trained and delegated to operate it, and shall not be started unless it is safe to do so. Personnel must not attempt to operate the compressor with a known unsafe condition. Operators and/or maintenance personnel shall tag the compressor and render it inoperative by disconnecting and locking out all power at its sources or otherwise disabling its prime mover so others, who may not know of the unsafe condition, cannot attempt to operate it until the condition is corrected.

- Only new oil from an oil manufacturer, not used or filtered oil, shall be added to a compressor.
- The compressor package should be leveled and securely anchored without base distortion.
- Checklists reviewed and approved by the Engineering, Operations, and Safety departments shall be used for all complex operations and maintenance procedures. The checklists shall list all steps in the operations of interest as well as the normal operating range (pressure, temperature, power consumption, etc.) for each step. This procedure permits early recognition of abnormal conditions.

Emergency Procedures

The purpose of this segment is to highlight the conceptual framework within which safety measures need to be included in an emergency preparedness plan for compressor operations.
- Facilities for emergency shutdown or isolation shall be provided for single compressor/multiple unit compressor systems when failure or operational deficiencies could create a serious hazard in adjacent areas. Whether such facilities should operate automatically or from a remote manual station will depend on the dangers involved. Valves for isolating each compressor and driver shall be provided.
- The arrangement of compressor and driver piping shall facilitate drainage. In addition, accessible separation and/or drainage facilities shall be provided in piping upstream of compressors or drivers. These facilities shall be at low points or other locations susceptible to collection of liquid. Compressor accessories such as water jackets and tubular heat exchangers shall be provided with drainage facilities to prevent freezing during idle periods.
- Screens should be installed in compressor suction during initial startup and break-in periods for protection against damage from foreign materials. The screens may be removed after the piping system has been cleaned. A pressure drop indicator is recommended.
- Reciprocating compressor cylinders handling saturated vapors should be furnished with discharge nozzles on the bottom or downward side to facilitate drainage when the unit is idle. Should, during an emergency, some gas enter the frame, relief valves shall be installed to provide with nitrogen purging to maintain safe operation.
- A check of emergency or safety trip devices on compressor equipment shall be included as part of regularly scheduled maintenance. Any safety device required by standards (ASME B19.3, ASME B19.1, also see Fig. 2 and Bibliography) shall not be jumpered, bypassed, or defeated by any person. In the event that shutdown of the system would impose a more hazardous condition, the owner shall install a backup system drive train to maintain safe operation of the overall plant.
- The emergency procedures to be used in correcting some abnormal conditions shall be also outlined in the checklist. The checklists containing normal and abnormal conditions shall be reviewed periodically as part of the safety program.

All operating personnel shall be trained in emergency response procedures.

Start-up and Shutdown

The startup, shutdown, and emergency procedures for all operations involving compressors shall be reviewed periodically with the operators. An understanding of the following general measures of accident prevention during startup and shutdown implementation process is thus important to operating personnel:

- Inlet piping to compressors or drivers, surge chambers, disengaging drums or scrubbers, and cylinders or casing shall be drained prior to startup.
- Whenever the compressor is to be shut down for service, warning tags on the electrical system shall be placed. Exposed electrical wiring must always carry a warning tag even though it is disconnected from the power supply.
- After mechanical work of any kind has been completed on a compressor, it shall be checked over sufficiently to ensure that there are no mechanical interferences within the compressor or driver.

Maintenance And Repair

An integrated compressor maintenance program may have any or all of the following actions:

- If the condition impairs safe operation, the compressor shall be taken out of service for repair in the prescribed safe manner.
- Safeguards that have been altered or damaged shall be reported so appropriate action can be taken to ensure against worker injury.
- The flywheel or crankshaft of a reciprocating compressor shall be locked in place prior to maintenance work. After mechanical work of any kind has been completed on a compressor, it shall be barred over sufficiently (at least one revolution) to ensure that there are no mechanical interferences within the compressor or driver.
- When maintenance is being performed on compressors, precautions shall be taken to ensure isolation of all energy sources to the driver. These precautions shall include the use of blinds or double valves and bleeder on steam and fuel gas suppliers to drivers, or in the case of motor driven compressors, either:
-electrical load centers shall have a switching arrangement that must be locked in the open position, tagged, and tried; or
-other positive means of current interruption shall be employed.
In all cases, all connected equipment shall be depressured to prevent rotation of the drive shaft.

SAFETY PRECAUTIONS AND HAZARD MITIGATION

The purpose of this section is to outline the conceptual framework within which safety measures and hazard recognition, mitigation and control can be implemented. An understanding of the principles of hazard protection and prevention helps the operating supervisor at the plant level (Fig. 1b) and safety and environmental specialists at the planning level (Fig. 1a) to determine what can realistically be applied for the safety program. Sets of corresponding standards and regulations, presented in Fig. 3, are recommended as being effective in the mitigation and control of different hazards associated with compressor operations.

Noise

Protection against the effects of noise exposure shall be provided in accordance with 29 CFR Part 1910.95. A compressor room noise level is estimated to be nominally 125 dB, which is in the range that can cause ear injury and progressive hearing loss. Compressor manufactures should be consulted regarding expected noise levels. Other system components such as valves, inlet piping, and separators can generate noise, and the manufactures of these components should be consulted also. Manufacturers are able to offer system components that have been specifically designed or treated to reduce their noise levels. Should a system not meet OSHA requirements, the affected area shall be clearly identified and warning signs posted at all entrances. Ear protection is required at all times when the facility is operating. Doors to such facilities shall be placarded with directions to wear ear protection. Personnel in the area wearing portable ODH monitors will not be able to hear audio warnings. Therefore, the compressor room shall also have the necessary fixed ODH monitors and visual alarms to alert personnel of ODH hazards or other hazards such as fire (Blyukher,1995).

Moving Parts

Moving parts are defined as those having movement that is not hand powered and which move during normal operation. Guards shall comply with ANSI/ASME B15.1.
All exposed moving parts shall be provided with personnel protection guards. These guards shall be designed and constructed so as to avoid risk of injury to personnel. Guards shall be:

- made of solid materials, with or without a liner, when necessary to meet the area classification (NFPA 70 and/or similar governing document.)
- securely fastened to the machine supporting structure or to the machines themselves.
- such that heat buildup or concentration of corrosive material will not adversely affect the parts being guarded or personnel.

Lubricants

When fire resistant or other synthetic lubricants are used, the following shall be considered:

- effect of lubricant on painted surfaces, gaskets, and seals;
- effects of process gas on lubricant in the cylinder;
- toxicity rating of lubricant;
- effects of lubricant carry-over into process equipment.

Spills of lubricant or other combustible substances shall be handled according to emergency response procedures, and reported to Safety immediately. The compressor shall be operated only in open or adequately ventilated areas.

Care should also be taken to avoid accidental ingestion and /or skin contact. In the event of ingestion, seek medical treatment promptly. Wash with soap and water in the event of skin contact.

Electrical Hazard

Electric motors, their controls, and wiring shall be installed and maintained in accordance with manufacturer's instructions and NFPA 70B. Should a conflict exist, the manufacturer's instructions should be given preference. The following safety requirements shall be implemented in accordance with NFPA 70E:
- before attempting to start the compressor, the electrical control system, protective switches and capacity controls must be checked in a simulated operating condition;
- the presence of oil in the separator shall be checked so the oil heater will not burn out;
- the electrical check, maintenance and repair must be made with the main motor disconnected;
- prior to attempting repairs or adjustments to rotating machinery and prior to leaving the compressor unattended with open electrical enclosures. all power at source shall be disconnected, locked and tagged out so others will not inadvertently restore power.

Hot surfaces, sharp edges and sharp corners

During compressor work hot oil, hot coolant, hot surfaces are generated and contact with them shall be avoided. Also, all parts of the body shall be kept away from all points of gas discharge and from sharp edges and corners. When working in, on or around the compressor, personal protective equipment including gloves and head covering must be worn to prevent burns and cuts.

Excess Pressure Protection and Pressure Release

Compressor safe performance in general is influenced by pressure release, gas leakage throughout the packing, and gas back flow across the valves due to overpressurization. The most frequent causes of overpressure on compressors are:
- blocked outlets or other restrictions to flow;
- failure of automatic controls;
- loss of cooling water;
- change in composition of gas or vapor;
- increase in suction or inlet pressure;
- excessive speed;
- flow reversal;
- the malfunction of reciprocating compressor valves.

The compressors system shall be analyzed to determine what circumstances or combinations thereof will cause the pressure on any compressor element to exceed 110% of its MAWP. For the most severe conditions, the flow at this pressure shall govern the capacity of relief facilities (Blyukher, 1992). All pressure relief devices and instruments shall be calibrated. The maximum pressure setting of relief devices shall adhere to the requirements of the ASME Code, Section VIII, Division 1, UG-131, 133, 134.

All relief valves are to be piped to an exhaust with sizing and locations per the Code requirements. Whenever working on any piping, tubing or other connections, compressors must be blown down to atmospheric pressure. Even when blown down, caution must be followed when loosening connections as localized pressure pockets can still be present.

Pressure Setting and Sizing of Relief Devices

The maximum pressure setting of relief devices shall adhere to the requirements of the ASME Code, Section VIII, Division 1, UG-134. To minimize leakage from pressure relief valves, the set pressure of valves (pressure at which relief valve starts to open) should be a minimum of 10% or 15 psi(103 kPa), whichever is greater, above the intended operating pressure at the valve inlet. On reciprocating compressors, the minimum should be 10% or 25 psi (172 kPa), whichever is greater.

Location of Pressure Relief Devices

For most compressor installations, overpressure protection for the compressor and its auxiliaries requires only a pressure relieving device or devices on the discharge of each compressor stage. Such an arrangement will usually suffice, provided the system pressure gradient under relieving conditions is such that the pressure on the weakest element will not exceed its maximum allowable working pressure by more than 10%.

Relief device should be installed as close as practicable to the system being protected. Pressure relief devices should be installed on the discharge and upstream of any check valves in the system.

Disposal of Relief Stream
- Atmospheric discharge pipes shall terminate at a location which will not create a hazard to personnel. Where feasible, direct relief to the outside atmosphere is recommended.
- Discharge from relief devices for equipment located in a building shall terminate outside the building, preferably above the highest point of the roof, for relief of gases other than air.
- Atmospheric discharge lines shall be designed to facilitate drainage of water or other liquids which may accumulate in the outlets of the relief devices on the discharge riser.
- When compressors are installed in a closed loop (see example of compressors at SSCL in "Introduction"), relief streams off the discharge may be routed to the suction portion of the system, but if disposal of relief stream, either to atmosphere or to a lower pressure portion of the system, is considered impractical or unsafe, relief streams should be discharged to a closed gathering system.
- Discharge lines of closed systems shall be designed to facilitate drainage of liquid which may accumulate in outlet piping from relief valves.

SUMMARY
1. This paper demonstrates how to establish and implement a Safe Compressor Operations Program by utilizing the basic principles of hazards protection and control utilized by OSHA and industries standards.
2. Primary potential hazards associated with compressor operations and facilities are: noise, heat burns, mechanical hazards associated with rotating equipment, remote possibility of oil or cryogen spill or leak, pressure, ODH, electrical hazards, and possibility of fire. These hazards are common for most compressor facilities, and the safety requirements and safety management program components discussed can be applied to all types of compressors (centrifugal, axial, rotary, and reciprocating) which are an integral part of facilities for processing petroleum, petrochemicals, or chemicals, including air separation plants.
3. Prior to installing or operating the compressor, users shall become familiar with, and comply with, all applicable OSHA regulations, Refrigerant and Oil Material Safety Data Sheets and applicable Federal, State and Local codes, standards, and regulations relative to personal protective equipment such as eye and face protective equipment, respiratory protective equipment, equipment intended to protect extremities, protective clothing, protective shields and barriers and electrical protective equipment, as well as administrative and/or engineering controls of noise exposure, and/or personal hearing protective equipment. Eye wash facilities must be easily reached when compressed gases are handled. These facilities must be routinely inspected and tested to ensure They are properly functioning and to change the water in the pipes.
4. Properly engineered closed systems, which may include air pollution control and monitoring equipment, must be provided to prevent gas from escaping into work areas and to prevent emissions in excess of each facilities ODH rating based on calculations in accordance with CGA SB-2, *Oxygen Deficient Atmospheres*, (see also (Blyukher, 1995). Fixed/Stationary oxygen monitors should be utilized for the purpose of early warning. These monitors shall be set to alarm for area evacuation at 19.5% oxygen.
A sign must be posted outside the work area to warn visitors and to provide information on emergency personnel notification.
5. A six level safety management program including corporate and procurement management, safety engineering, specifications and requirements, and operations and production is proposed for each compressor system. It is recommended to incorporate this program into Occupational Safety and Health programs and Emergency Preparedness plans. Through establishment and implementation of a compressor hazards prevention and mitigation program, risk to the work place and work force from the release compressed gases, pressure and ODH hazards is greatly reduced.

BIBLIOGRAPHY

29 CFR, 1910. 101, Compressed Gases (general requirements), 1910.219, Mechanical Power-Transmission Apparatus.

29 CFR, 1926, Subpart V, Power Transmission and Distribution.

DOE 6430.1A *"General Design Criteria"*

ASME Boiler and Pressure Vessel Code, Section V111, 1992 Edition.

NEMA, MG2, Safety Standard for Construction and Guide for Selection, Installation, and Use of Electric Motors and Generators.

ASME PTC 9-1970, March 1986, Displacement Compressors, Vacuum pumps and Blowers.

ASME B15.1, Safety Standard for Mechanical Power Transmission Apparatus.

ASME B 19.3-1991 "Safety Standards for Compressors for Process Industries".

ASME B 19.1-1990 "Safety Standards for Air Compressors Systems".

ARI Standard 530-89 "Methods of Measuring Sound and Vibration of refrigerant Compressors", Air-Conditioning & Refrigeration Institute.

API Recommended Practice 520, "Sizing, Selection, and Installation of Pressure-Relieving Devices in Refineries", 6th edition, March 1993, American Petroleum Institute.

API Recommended Practice 500 (RP 500), "Recommended Practice for Classification of Locations for Electrical Installations at Petroleum Facilities", 1st edition, June 1991, American Petroleum Institute.

API Standard 618, "Reciprocating Compressors for General Refinery Services", 3rd edition, February 1986, American Petroleum Institute.

ASME B 31.3-1993 "Chemical Plant and Petroleum Refinery Piping".

CGA SB-2, *Oxygen Deficient Atmospheres*, Compressed Gas Association, 1987.

Blyukher, B., 1995, Oxygen Deficiency Hazard Analysis for Pressure and Cryogenic System Facilities,"*Structural Integrity of Pressure Vessels, Piping, and Components*" , *PVP-Vol.318*, ASME/JSME Pressure Vessels and Piping Conference, July 23-27 1995, Honolulu, Hawaii, pp. 169 - 175.

Blyukher, B., 1992, " Refrigeration Plant Compressor System Safety", Technical Procedure AQA-1010045, Superconducting Super Collider Laboratory, Dallas, Texas.

ACKNOWLEDGMENTS

In compressor safe operations and corresponding standards overview presented in this paper the conceptual approaches and procedures used at SSCL, Cryogenic Co. "KISLORODMASH" and other cryogenic, and chemical companies in Russia and Ukraine have been utilized. I would like to thank the key members of this "Compressor Safety" team for their input to this program. They include George Mulholand, Harry Carter, John Urbin, David Hawkins, Chuck Watt, and Clayton Hamm. I would also like to thank ASD Director Dr. Ted Kozman and ASD Safety Manager Ray Nations for their direction and constructive input on this program. I would like to extend my appreciation to Dr. Lev Reznikov for his input in observation of safety measures for reciprocating compressors operations.

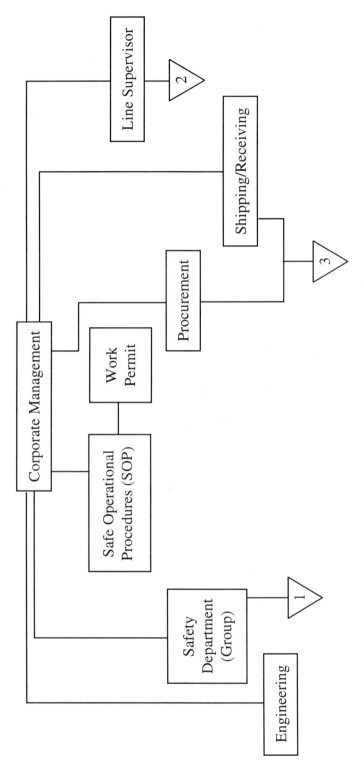

Figure 1. Safe Compressor Operations: Levels of Organization

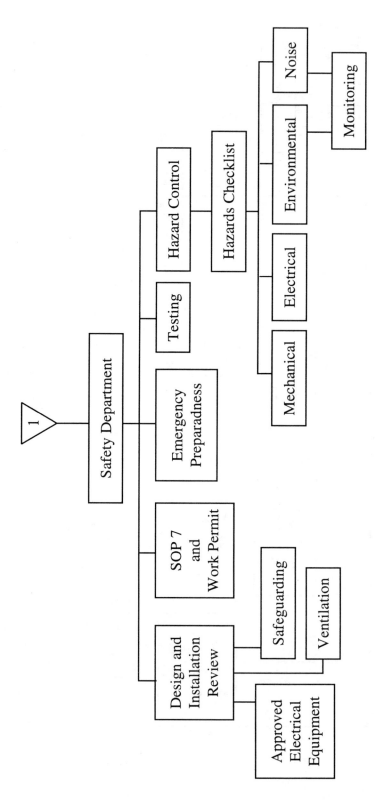

Figure 1a. Safe Compressor Operations: Level of Organization (con't)

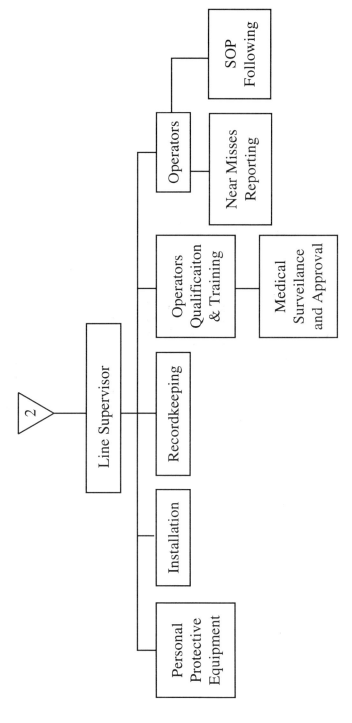

Figure 1b. Safe Compressor Operations: Level of Organization (con't)

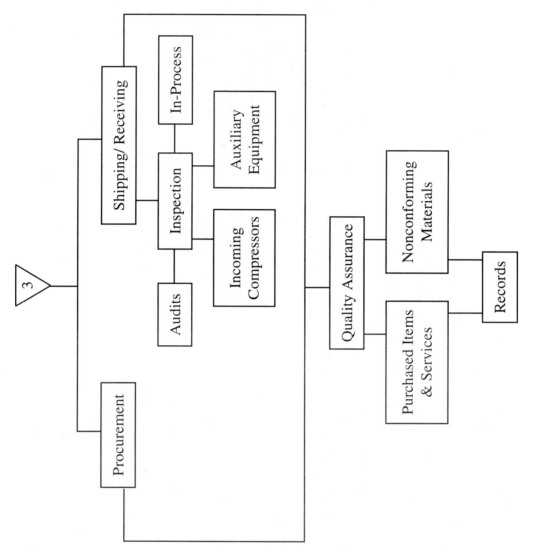

Figure 1c. Safe Compressor Operations: Level of Organization (con't)

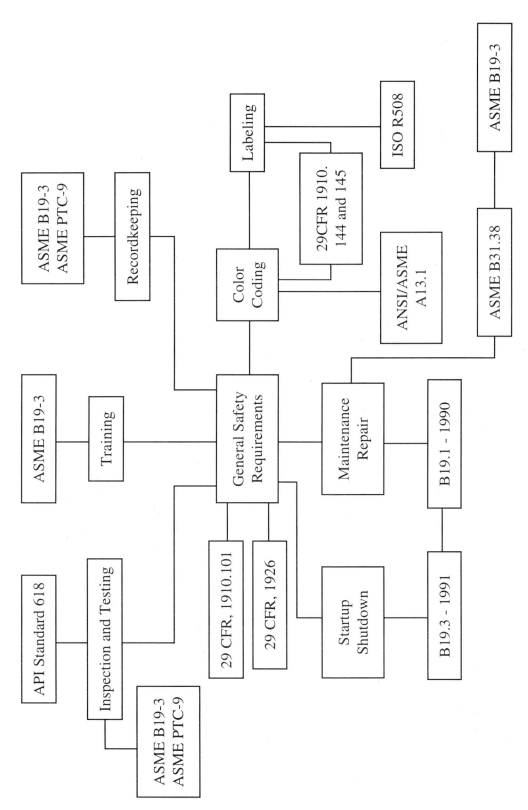

Figure 2. Set of standard and regulations covering general safety requirements for safe compressor operations.

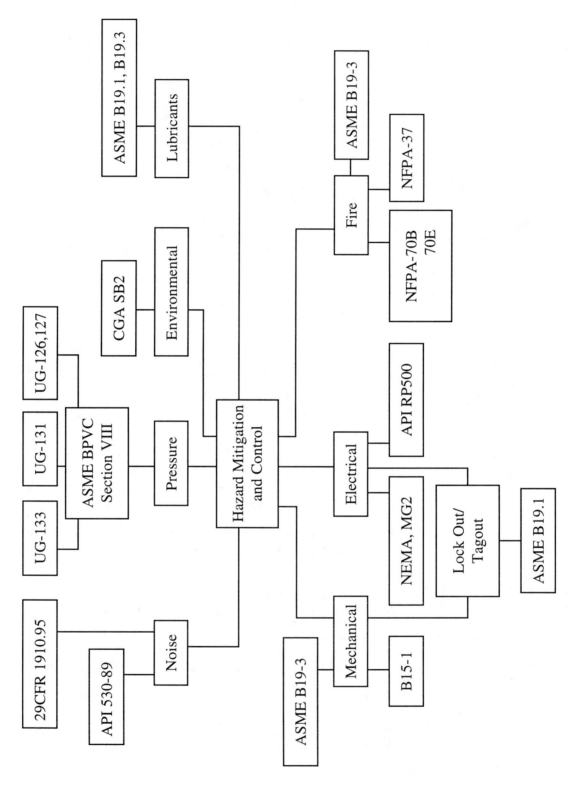

Figure 3. Set of standard and regulations covering hazard mitigation and control for compressor operation and maintenance.

PIPE SUPPORTS AND RESTRAINTS AND RELATED TOPICS

Introduction

Evans C. Goodling, Jr.
Parsons Power Group, Inc.
Reading, Pennsylvania

Four papers in this session address a variety of piping issues involving support and restraint of piping in power plants and other industrial facilities.

Professor *Smith* extends his earlier work on the behavior of restrained piping with circumferential cracks, focusing on the study of crack-growth instability as a result of pipe restraint failure. *Azzazy, Gateley, Ward and Watts* discuss a wide variety of piping structural integrity problems encountered in aging power plants, including the role of computer-based analysis in understanding and resolving these problems. *Goodling* offers a commentary on the buried piping analysis procedures cited in the ASME B31.1 Power Piping Code with emphasis on accurate modeling of the non-linear passive restraint effects of compacted soil backfill on the pipe. *Robleto* describes the design of a complex restraint on a large diameter pipe in order to provide moment restraint while permitting necessary translational displacement of the pipe.

**PVP-Vol. 356, Integrity of Structures and Fluid Systems,
Piping and Pipe Supports, and Pumps and Valves
ASME 1997**

THE EFFECT OF RESTRAINT FAILURES ON UNSTABLE
FRACTURE OF A PIPING SYSTEM

E. Smith

Manchester University - UMIST Materials Science Centre,
Grosvenor Street, Manchester M1 7HS, United Kingdom.

ABSTRACT

During the last two decades, considerable attention has been given to the structural integrity of steel piping systems, and in particular to the effect of circumferential cracks on their integrity. From a safety perspective, it is important that any crack, say for example a stress corrosion crack, will not lead to a guillotine rupture and thus possibly a pipe whip scenario. One way of guaranteeing that this does not happen is to ensure that unstable growth of a circumferential crack is unable to occur, the appropriate methodology being based on tearing modulus concepts. A key element in this methodology is the crack-system compliance length parameter L_*, with crack instability being more likely with a high L_* value. It is against this background, and on the basis of a simple model analysis, that the paper highlights the effect of restraint failures, for example snubber failures, on increasing the value of L_*, and thereby encouraging instability of crack growth.

INTRODUCTION

Interest in the integrity of cracked piping systems fabricated from materials that are generally accepted as being ductile, has been motivated, to a considerable extent, by the technological problem of intergranular stress corrosion cracking (IGSCC) of Type 304 stainless steel piping in Boiling Water Nuclear Reactor piping systems. IGSCC cracks have been found at the inner surface of pipes, and are usually associated with the heat affected zones of girth welds. The cracks are circumferential and are able to grow slowly in service by a time dependent environmentally assisted mechanism. From a safety perspective, it is important to know whether accident condition loadings, due for example to an earthquake, will allow a part-through stress corrosion crack to propagate unstably across the pipe thickness by a non-environmentally assisted (plastic) fracture mechanism, and the resulting through-wall crack then propagate around the pipe circumference and cause complete pipe severance, thus possibly leading to a pipe whip scenario. This particular problem has been the focus of considerable attention, against the background of the IGSCC problem, but the issue of unstable circumferential crack growth and its potentially serious consequences has a much wider range of significance where piping systems are concerned.

Extension of a crack can be predicted by assuming that extention conforms to a net-section stress criterion[1], which uses as input an appropriate value for the critical net-section stress, usually the average of the material's yield and ultimate tensile stresses, together with a knowledge of the anticipated loadings. The stress at the cracked section, which is compared with the critical net-section stress, is usually calculated via a purely elastic analysis based on the piping being uncracked.

However, because the piping system is built-in at its ends into a larger component, for example a pressure vessel or steam generator, and because the onset of crack extension requires some plastic deformation, use of the net-section approach for predicting the onset of crack extension can give overly conservative failure predictions. The author recognized[2] the importance of this effect and has presented papers at recent ASME PVP Conferences which quantify this conservatism in terms of the geometry of the cracked section, system geometry and material ductility.

Even taking into account the aforementioned conservatism, it can be argued that because accident loadings cannot be estimated with any reasonable accuracy, an alternative and more conservative approach should be employed aimed at preventing crack extension becoming unstable. The foundations of the appropriate methodology were developed for piping in the pioneering paper by Tada, Paris and Gamble[3], based on the tearing modulus procedures formulated by Paris and co-workers[4]. A key element in this methodology is the crack-system compliance length parameter L_*, with crack instability being more likely with a high L_* value, an extremely crude measure of which is the pipe-run length.

It is against this background, and on the basis of a simple model analysis, that the present paper highlights the effect of support failures, for example the inoperability of snubbers during an accident loading scenario, on increasing the value of L_*, and thereby encouraging the instability of crack growth.

FORMULATION OF THE GENERALIZED INSTABILITY CRITERION

The cracked cross-section of a circular cylindrical pipe, thickness t and radius $R >> t$, is shown in Figure 1; the cross-section contains a through-wall crack with contained angle 2θ. So as to simplify the considerations, it is assumed, following Tada, Paris and Gamble[3] that: (a) the material does not work-harden, (b) the material is sufficiently ductile that the onset of crack extension occurs after general yield, i.e. the "Battelle Screening Criterion"[5] is satisfied, (c) the deformation at the cracked section can be represented by a rotation about the neutral axis. With the cracked section being subjected to the plastic limit moment M_p, the stress distribution across the section can be described by a tensile stress σ_o acting within the region above the neutral axis, while a compressive stress of similar magnitude is operative below the neutral axis. Assuming that the effect of any resultant tensile force acting perpendicular to the cracked section can be neglected, the location of the neutral axis, defined by the angle α, is readily shown by balancing forces, to be given by $\alpha = \theta/2$, whereupon moment equilibrium provides the relation

$$\frac{M_p}{4\sigma_o R^2 t} = \cos\alpha - \frac{1}{2}\sin\theta = \cos\frac{\theta}{2} - \frac{1}{2}\sin\theta \quad (1)$$

The crack tip opening displacement D_{TIP} is given, via simple trigonometrical arguments, by the expression

$$D_{TIP} = R(\sin\alpha + \cos\theta)\phi = R\left[\sin\frac{\theta}{2} + \cos\theta\right]\phi \quad (2)$$

where ϕ is the rotation about the neutral axis. Thus with the J integral expressed as the product of D_{TIP} and the yield stress σ_o, it follows that

$$J = \sigma_o R\left[\sin\frac{\theta}{2} + \cos\theta\right]\phi \quad (3)$$

The tearing modulus methodology for predicting crack instability, due to Paris and co-workers[4], is based on the premise that the instability criterion is $T_{APP} > T_{MAT}$ where T_{MAT} is the material tearing modulus, which is related to the initial slope of the J-crack growth resistance curve in an appropriate experimental test:

$$T_{MAT} = \frac{E}{\sigma_o^2} \frac{dJ}{da} \qquad (4)$$

where E is Young's modulus. There is a corresponding expression for T_{APP}, the applied tearing modulus, and this can be obtained from relation (3) noting that $\delta a = R\delta\theta$ for circumferential crack growth. Thus if J_{IC} is the value of J at the onset of crack extension, the instability criterion is

$$T_{APP} = \frac{EJ_{IC}}{\sigma_o^2 R} \cdot \frac{\left[\frac{1}{2}\cos\frac{\theta}{2} - \sin\theta\right]}{\left[\sin\frac{\theta}{2} + \cos\theta\right]}$$

$$+ \frac{E}{\sigma_o}\left[\sin\frac{\theta}{2} + \cos\theta\right]\frac{d\phi}{d\theta} > T_{MAT} \qquad (5)$$

Now, following Tada, Paris and Gamble[3], with a very ductile material that has a high resistance to plastic crack growth, i.e. a high T_{MAT} value, the term involving J_{IC} in expression (5) for T_{APP} can be neglected, and then the instability criterion can be expressed in the simplified form

$$T_{APP} = \frac{E}{\sigma_o}\left[\sin\frac{\theta}{2} + \cos\theta\right]\frac{d\phi}{d\theta} > T_{MAT} \qquad (6)$$

Now the actual moment M_a at a cracked section and the moment M_u, calculated on the basis that the piping system is uncracked, are related by an expression of the form

$$M_u = M_a + \frac{EI\phi}{L_*} \qquad (7)$$

where $I = \pi R^3 t$ is the second moment of area of the piping at the cracked section and L_* is a length parameter, which can be viewed as a crack-system compliance length parameter. With the applied loadings fixed, i.e. M_u is constant, and with $M_a = M_p$, relations (1), (6) and (7) give the crack instability criterion as

$$T_{APP} = \frac{2L_*}{\pi R}\left[\sin\frac{\theta}{2} + \cos\theta\right]^2 > T_{MAT} \qquad (8)$$

The form of relation (7) allows the crack-system compliance length parameter L_* to be obtained via a very simple procedure[6]. This involves a separation of the complete piping system into two elastic parts at the cracked section, the application of equal and opposite moments M' to the cut faces, and the correlation of M' with the rotational discontinuity ϕ' generated at this section, i.e.

$$L_* = \frac{EI\phi'}{M'} \qquad (9)$$

For a simple straight pipe segment, built-in at its ends, and containing a cracked section at the mid-length position, $L_* = L$, the pipe length, when the ends are subjected to applied rotations or displacements[3].

Relation (8) shows that a key element in determining whether crack growth is unstable is the crack-system compliance length parameter L_*, with crack instability being more likely with a high L_* value. In the next section, we use the results from the analysis of a very simple model to highlight the point that if supports, for example snubbers, fail during an accident load scenario, then L_* increases, thereby leading to a greater tendency for crack growth to become unstable.

THE EFFECT OF SUPPORT FAILURE ON CRACK INSTABILITY

The particular simple model that will be used to highlight the effect of support failure on the instability of circumferential crack growth is shown in Figure 2. A pipe of length L is built-in at both its ends, which are subjected to applied rotations and displacements which simulate the accident loadings. There is a cracked section, like that shown in Figure 1, situated at a distance s from the left-hand built-in end while there is a restraint situated at a distance $h(>s)$ from the left-hand end, and this prevents the pipe from moving in a vertical direction, it being assumed, for sake of simplicity, that the restraint acts in a rigid manner. The crack is assumed to be in that region (upper or lower) of the pipe cross-section such that it opens under the applied loadings.

For this particular model, it has been shown[7] that the crack-system compliance length parameter $L_*(R)$, with the restraint assumed to be fully operable, is given by the expression

$$\frac{L*(R)}{L} = \frac{h^3}{3L(h-s)^2 + h(3s-h)^2} \quad (10)$$

or with $s/h = x < 1$ and $h/L = \lambda < 1$

$$\frac{L_*(R)}{L} = \frac{\lambda}{(3+\lambda) - 6x(1+\lambda) + 3(1+3\lambda)x^2} \quad (11)$$

On the other hand, for the case where the restraint is inoperable to the extent that there is no restraint, then the crack-system compliance length parameter $L_*(NR)$ for this situation, is given by the expression

$$\frac{L_*(NR)}{L} = \frac{L^2}{4(L^2 - 3sL + 3s^2)} \quad (12)$$

or again with $s/h = x < 1$ and $h/L = \lambda < 1$

$$\frac{L_*(NR)}{L} = \frac{1}{4(1 - 3\lambda x + 3\lambda^2 x^2)} \quad (13)$$

It follows from relations (11) and (13) that the ratio of the two crack-system compliance length parameters is

$$\frac{L_*(NR)}{L_*(R)} = \frac{(3+\lambda) - 6x(1+\lambda) + 3(1+3\lambda)x^2}{4\lambda(1 - 3\lambda x + 3\lambda^2 x^2)} \quad (14)$$

Inspection of relation (14) clearly highlights the point that inoperability or removal of a restraint increases the crack-system compliance length parameter L_* thereby making crack instability more likely, since $L_*(NR) > L_*(R)$ for a cracked section irrespective of its location or the restraint's position. To illustrate the magnitude of the effect, consider the case where the restraint is at the mid-length position of the pipe when $\lambda = h/L = 0.5$. Relation (14) then becomes

$$\frac{L_*(NR)}{L_*(R)} = \frac{7 - 18x + 15x^2}{4 - 6x + 3x^2} \quad (15)$$

whereupon it is seen that $L_*(NR)/L_*(R)$ has the values of 1.75, 1.56, 1.37, 1.19, 1.06 and 1.00 for respectively $s/h = 0$ (crack at built-in end), 0.1, 0.2, 0.3, 0.4 and 0.5 (cracked section and restraint position coincide). It is immediately seen that inoperability or removal of a restraint can have a very large effect on L_*, e.g. an increase by a factor of 1.75, for the case where the cracked section is at the built-in end.

DISCUSSION

Recognizing that the applied tearing modulus $T_{APP} \alpha L_*$ (see relation (8)), the results obtained from the simple model considerations in the preceding section have clearly highlighted the beneficial effects that restraints can have on the stability of circumferential crack growth. It has been clearly demonstrated that inoperability or removal of a restraint during an accident loading scenario can lead to a large increase in the crack-system compliance length parameter L_*, and thus encourage the instability of circumferential crack growth.

Expressed another way, the results support the view that an upper bound for L_* for a piping system, which is associated with the most conservative perspective, can be obtained by assuming there are no restraints, which of course has the added advantage of simplifying the analysis. Alternatively, an analysis conducted on this basis allows for the fact that restraints might become ineffective during an accident.

CONCLUSION

● A simple model analysis has highlighted the effect of restraint failures on increasing the value of the crack-system compliance length parameter, and thereby encouraging the instability of growth of a circumferential through-wall crack in a piping system.

REFERENCES

1. American Society of Mechanical Engineers, Boiler and Pressure Vessel Code, Section XI, IWB-3640, Winter Addenda (1983).
2. E. Smith, Int. Jnl. Press. Vess. and Piping, 52 (1992) 257.
3. H. Tada, P.C. Paris and R.M. Gamble, ASTM STP 700 (1980) 296.
4. P.C. Paris, H. Tada, A. Zahoor and H. Ernst, ASTM STP 668 (1979) 5.
5. G.M. Wilkowski, Nuc.Eng. and Design, 98 (1987) 195.
6. K.H. Cotter, H.Y. Chang and A. Zahoor, EPRI Research Project T118-9, NP-2261, Final Report, Electric Power Research Institute, Palo Alto, CA, USA (1982).
7. E. Smith, PVP-Vol. 103, Proceedings of ASME Conference on Fatigue and Fracture Assessment, Chicago, Illinois, USA, (1986) 55.

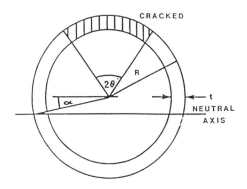

Figure 1. The geometry of the cracked section.

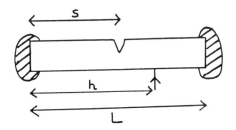

Figure 2. The model analysed.

PVP-Vol. 356, Integrity of Structures and Fluid Systems,
Piping and Pipe Supports, and Pumps and Valves
ASME 1997

APPLICATION OF STATIC AND DYNAMIC PIPING ANALYSIS AS AN AID TO UNDERSTANDING AND RESOLVING FIELD PERFORMANCE PROBLEMS

Salah E. Azzazy
Royce A. Gateley
Charles P. Ward
Larry H. Watts

Tennessee Valley Authority
Fossil & Hydro Power
1101 Market Street
Chattanooga, Tennessee 37402
USA

ABSTRACT

The Tennessee Valley Authority operates one of the oldest fleets of coal-fired power plants with 34 of the 59 units (at eleven plants) having operated in excess of 39 years. The piping systems in most of the units were designed prior to the advent of computer software for either static or dynamic analyses. Indeed, most of the units were designed without formal consideration of resistance to dynamic events such as turbine trips, safety valve lifts, or water/steam hammer events. Nevertheless, the performance of the piping systems has generally been good. This paper illustrates five different categories of operational problems that have been encountered with piping systems and the role of computer-based analyses in understanding and resolving these problems.

INTRODUCTION

Many of the old operating coal-fired power plants were designed and built prior of the development of computer analysis technology. In the area of piping systems, the lack of computer analysis records is a painful reality we have to face when conducting a failure/root cause analysis to find a solution to an operational problem. With the current availability of a wide range of computer software, especially for all forms of static and dynamic analyses, the task of finding the cause and solution of an operational problem has become easily achievable. This paper presents five categories of piping related operational problems that have occurred at different plant sites. The paper also shows the usefulness of computer-aided analysis as a tool to simulate the mishap event and to confirm a possible solutions. These categories are as follows:

1. Operational problems of rotating equipment attributed to excessive loading imposed by the piping system. Examples include boiler feed pumps and steam turbines.
2. Resonance that occurs during specific periods of unit operation. Examples include deaerator piping and startup piping.
3. Massive hanger and structural attachment damage due to shock event. Example is a cold reheat piping system.
4. Long term pipe movement (drift). Examples are main steam and hot reheat piping in which the hangers are almost exclusively constant load types.
5. Girth/saddle weld cracking in high temperature (main steam and hot reheat) piping due to inadequate flexibility or geometric configuration.

The selected categories cover a variety of problem events and methods utilized to develop solutions to these events. The paper emphasizes the role of static and dynamic piping analyses in complementing non-destructive examination, metallurgical analysis, and field inspection of piping and pipe supports as a failure-analysis/problem-resolution tool. The paper also emphasizes the importance of the initial step of data gathering in form of event chronology, design information, operation history, field inspection, and previous or similar events to build a strategy prior to the use of computer analysis technique.

CATEGORY 1
EVENT #1

This section illustrates operational problems of rotating equipment (boiler feed pumps) attributed to excessive loading imposed by the piping system. A turbine driven boiler feedwater pump wrecked within weeks of its initial operation.

BACKGROUND

Bull Run Fossil Plant is a stand alone 900-MW pulverized coal-fired unit that went commercial in 1965. It is a supercritical pressure unit employing an ABB-CE twin divided furnace, GE cross-compound turbine and two turbine-driven Worthington boiler feed pumps. The unit has been recognized world-wide for its outstanding heat rate on numerous occasions.

From initial operation up until the early 1990's, the unit has had a history of pump wrecks occurring approximately every 18 months. With this operating history and the recent technological advances in pump design, it was believed that a new pump "B" (Fig. 1.1.1) would be the solution. Unfortunately, this was not the case; as the new pump "B" also wrecked within eight weeks of initial operation. After making repairs and returning to service, the pump again wrecked within a matter of weeks. Since the pump was under warranty, a joint team (Vendor representative and TVA Bull Run staff) assembled to investigate the situation. Optical micrometers were used to "watch" the movement of the pump as it heated. The pump casing not only moved up but it also rotated slightly (neither should have occurred). A type "F" (floor mounted) variable spring hanger on the bottom pump discharge line was found binding. Therefore, it was acting as a rigid support. Whereupon, TVA central engineering was summoned to provide assistance in finding the solution.

Engineering soon discovered that the plant had occasional vibration problems in each of the pumps. The vibration seemed to be induced by the other pump with the common suction header serving to transmit the vibration (Fig 1.1.1).

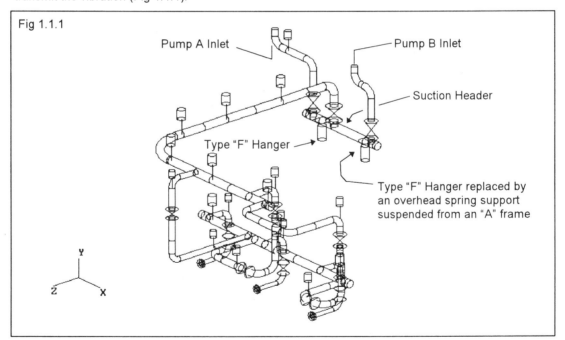

Fig 1.1.1

Pump A Inlet — Pump B Inlet

Suction Header

Type "F" Hanger →

Type "F" Hanger replaced by an overhead spring support suspended from an "A" frame

SOLUTION

The plant staff requested the "bound" hanger be replaced with a different type support. Also, they requested that the unit be kept on line during the repair. A "fix" and installation plan was devised and work began around the clock until the bound hanger was successfully replaced by an overhead spring support suspended from the turbine foundation. Following the repair, pump "B" was slowly brought up to operating temperature while the vender rep closely surveyed the movements of both pump housings. There was some improvement, but both pumps still had movements which seemed to parallel each other. Central engineering then concluded that the common suction piping was causing this problem. Inspection of the header (see Fig. 1.1.1) revealed: two "type F" hangers near each end indicating that they also would bind as the header expanded. Another similar fix was designed and installed. Again, some improvement was detected, but it was not a complete solution.

The main problem was determined to be lack of sufficient straight pipe supplying each pump with the resulting unbalanced hydraulic forces on the pumps. The inlet line from the common header to each pump was too short and stiff to allow sufficient vibration dampening between the pumps. This determination helped explain why certain operating speeds and pump operating combinations were worse than others. Tight space limitations in this area did not allow for lengthening and "straightening" of the inlet piping.

Coincidentally, this inlet line was slated for replacement due to wear (flow assisted corrosion). Therefore, engineering had freedom to reroute the pipe to resolve the pump problems. The problem was ultimately solved by removing the header entirely and installing a wye block on the main supply line with a large "U" shaped loop to each pump (Fig. 1.1.2). Supports and guides were installed on these loops to provide the needed stability and isolation for long-term operation.

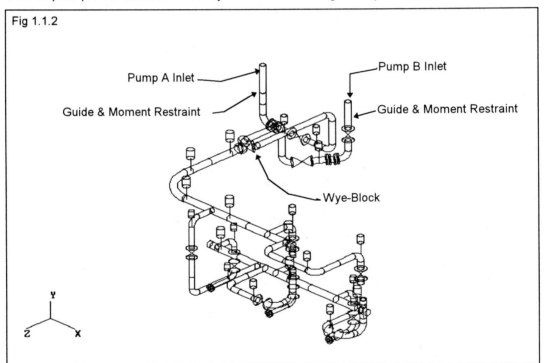

Fig 1.1.2

CONCLUSION
During the next maintenance outage, not only was the suction piping replaced, Pump "A" was replaced. These boiler feedwater pumps have now run over four years without wrecking.

CATEGORY 1
EVENT #2

This section illustrates operational problems of rotating equipment (steam turbines) attributed to excessive loading imposed by the piping system. Shawnee Fossil Plant Unit 6 (190 megawatt unit) was tripped due to high turbine vibration.

BACKGROUND

DATE	TIME	COMMENTS
2-9-96		While High Energy Piping (HEP) personnel were performing a "Quick-Look" of hangers, two adjacent hangers on the east lead of the high temperature reheat were found with broken U-bolts and reported to the plant.
2-11-96	1000	Checking out unit, preparing to return to service after being on cold stand-by
	1330	Unit operator heard loud metallic bang-Assistant Unit Operator was sent to investigate (suspected a pulverizer)
	2110	Unit on line
2-12-96	1715	Trouble adjusting turbine seal steams
	1745	Encountering bearing vibration in #4 turbine bearing
	2100	Bearing vibration still cycling
	2230	Reducing load on unit, trying several different things to slow vibration, hooking up external vibration monitor to check vibration.
	2235	Can hear noise in IP & LP casing.
2-13-96	0400	Vibration on #4 bearing at 10 mils, nothing seems to help, unit tripped due to high turbine vibration
2-14-96		Turbine is opened to determine damage.
2-16-96		HEP personnel back in area to determine if hanger had been repaired, a third hanger nearer the turbine had been pulled from its traveler. This situation was reported to plant staff with the recommendation not to start the unit before repairs could be made to the hangers.
2-20-96		Turbine repairs continue, preparation for hanger repair starts.
2-24-96		All turbine and **hanger** repairs completed and preparation for restart begun.
2-25-96		Unit on line, no problems

ROOT CAUSE

When the turbine was opened, a shroud was found missing on a series of blades which caused the turbine vibration. The rotor had rubbed the shell to the extent that the retainer tenon wore off. Apparently, the shell had been distorted sufficiently due to the broken hangers to make contact with the rotor. To determine what forces had been imposed on the turbine to cause this distortion, HEP performed a standard piping analysis (HR6022) to establish basis values of forces and moments that should be on the turbine. An additional analysis (HR7032) was performed with the three broken hangers (490, 510, & 520 location shown in Fig 1.2.1) removed.

The table below lists the forces and moments on the turbine shell and the calculated resultant for the forces and moments. Node 1750 is the east lead connection to the turbine shell, and Node 2050 is the west lead connection to the turbine shell. Allowable values are given by Westinghouse on the general arrangement drawing 1-J-3174: Resultant Force ≤ 2000 (lb.) and Resultant Moment ≤ 16667 (ft.-lb.).

Job number	Node	Force (lb.)					Moment (ft.-lb.)				
		FX	FY	FZ	RF	% Allowable	MX	MY	MZ	RM	% Allowable
HR6022 basis	1750	-774	-879	-25	1171	59%	10884	-10764	-29251	33014	198%
HR7032 broken	1750	1119	-5942	-50	6047	302%	94749	16676	-1E+05	144688	868%
HR6022 basis	2050	-730	-872	18	1137	57%	-10711	10296	-28698	32316	194%
HR7032 broken	2050	-1151	-920	-54	1474	74%	-11295	17919	-33384	39537	237%

Note that the resultant forces are less than the allowable except for the run with the broken hangers on the east lead. Whereas, the resultant moments exceed the allowable in all cases.

Plant personnel observed a large sag in the east leg and requested a theoretical value be determined. Analysis (HR7032) with three hangers broken, showed, as one would expect, significant displacement. The maximum deflection of the pipe was 6 ¾ inches.

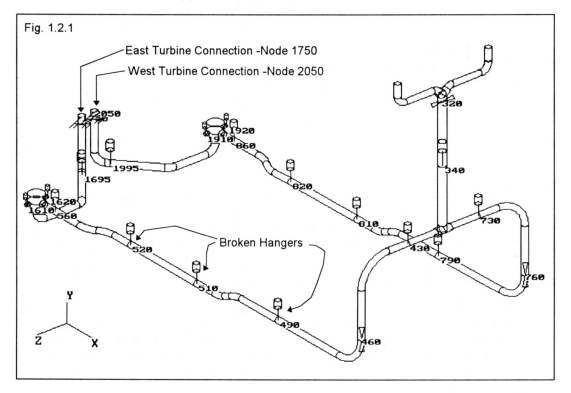

Fig. 1.2.1

SOLUTION
Repair the broken hangers. The broken U-bolts were analyzed and found to be the wrong material for the service temperature. A program was initiated to replace existing u-bolts with u-bolts made of ASTM A-479 type 316H stainless steel.

CATEGORY 2

This section illustrates operational problems with resonance in deaerator piping that occur during specific periods of unit operation.

EVENT #1

The weld of the outlet from the deaerator tank cracked. The water supply to the boiler feedwater pumps was lost and the unit was forced off line. Only after the third of five sister units at the TVA Kingston Fossil Plant experienced a similar failure was the High Energy Piping Group called to identify the root cause of the failure and develop a solution.

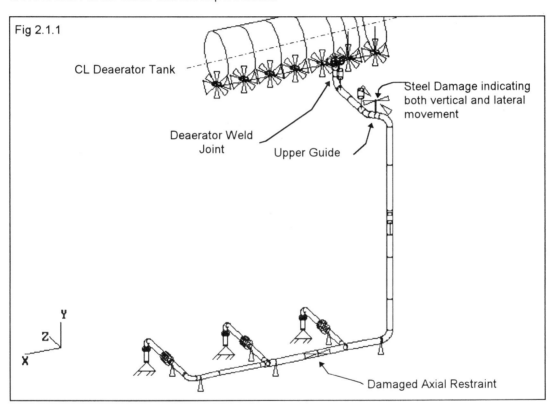

Fig 2.1.1

CL Deaerator Tank

Steel Damage indicating both vertical and lateral movement

Deaerator Weld Joint

Upper Guide

Damaged Axial Restraint

INVESTIGATION

Step 1. Standard procedure is to perform a walkdown of the affected units and sister units to identify broken or damaged hangers and insulation damage evidencing excessive pipe movement. Results of the walkdown indicate that no hangers appeared broken or seriously damaged. Vertical and lateral movement of the upper guide, which was a 4 inch pipe welded to the 16" and inserted in a slot in an 8" channel, had damaged the area around the control slot. The axial restraint on the 16" pump suction header appeared damaged (Fig. 2.1.1) and insulation damage was minimal. The pipe could be moved by hand with very little force.

Step 2. A computer model of the piping system including the deaerator tank and the pump connection was developed to analyze stress at the tank-pipe connection. A standard analysis was conducted to confirm that the piping conforms to ANSI B31.1 stress allowable and that hanger sizes are reasonable. The results show that the sustained and thermal stresses were very low. However, a large bending moment about the x-axis (Fx = -19 lb., Fy = 389 lb., Fz = 284 lb., **Mx = 14565 ft.-lb.**, My = 171 ft.-lb., Mz = -4515 ft.-lb.) on the weld was identified.

Step 3. A dynamic analysis was required since no other problems were found. The initial step of the dynamic analysis is solution of natural frequencies. The first natural frequency (Fn) was very low indicating that a small forcing function could cause the pipe to oscillate. The most likely source of a forcing function was the variable flow rate. Unfortunately, the flow rate could not be stabilized to eliminate the problem because the unit is controlled by varying the flow rate. Therefore, the pipe motion had to be controlled.

SOLUTION
Add rigid restraints at convenient locations and rerun thermal, static, and dynamic analyses to check for an improvement in system response.
   A. Adding a restraint near the weld would convert the bending moment to a force (Fx = 178 lb., Fy = 461 lb., Fz = 3874 lb., **Mx = 1720 ft.-lb.**, My = -2971 ft.-lb., Mz = -4186 ft.-lb.). Although, the resultant force increased 810%, the resultant moment decreased 281% and the stress was reduced 29.4%.
   B. Adding restraints at strategic points did not impair the thermal expansion, but continued to raise the first Fn to larger values (from 2.65 Hz to 10.9Hz) with each attempt.
   C. The final modification included: two restraints to limit the lateral movement, a restraint on the piping to convert the bending moment to a force, and a new stronger axial restraint in the pump suction header (Fig 2.1.2).

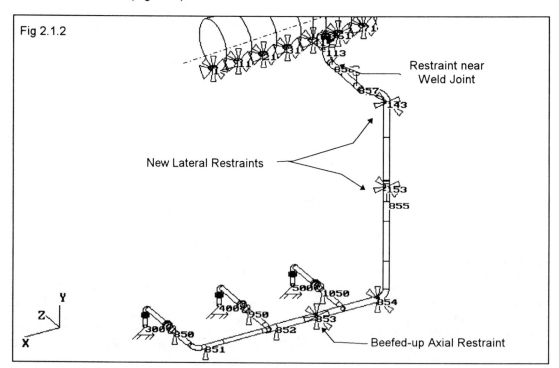

Fig 2.1.2

CONCLUSION
At the next maintenance outage, each unit had the restraints installed and the improvement in the system operation was dramatic.

CATEGORY 3

This section illustrates massive hanger and structural damage due to shock events.

EVENT

An "obvious" waterhammer in a cold reheat line which resulted in twelve broken hangers and pipe sagging.

BACKGROUND

Kingston Fossil Plant Unit 6 was being brought on line after a maintenance outage. The turbine had undergone repairs and a balance shot would be required. The turbine had been spun to obtain vibration data. The turbine was returned to turning gear. Calculations were being made to size the balance shot, craftsmen were repairing a leaking thermowell, and the boiler continued to be fired. The repairs were made and the balance shot installed. Preparations were made to spin the turbine. The main steam control valves were opened. The turbine spun easily with little valve opening, but soon slowed. The operators continued to open the valves to increase turbine speed. Suddenly, turbine speed shot up and a large crash was heard in the control room. An Assistant Unit Operator (AUO) was dispatched to determine the source of the noise. A pulverizer was suspected, as there is a history of "puffs". As he left the control room, he found hangers on the floor. The turbine was returned to turning gear, and the boiler shut down.

At 10:30 p.m. on a Saturday night, the High Energy Piping group was called to the site to help determine the cause and assist with repairs.

DAMAGE EXTENT

Twelve of the twenty-two hangers on the cold reheat system (see Fig 3.1.1) were broken in some fashion, and two others stretched to the maximum. Many were in the floor, however three or four had failed but were still holding the pipe in the air. Hanger support steel (back-to-back channel) had been bent and the clip angles ripped from the main beams, pipe clamps had been stretched, and grating at vertical penetrations had been crumpled under the excessive force. The damage was almost a duplicate on each leg of the cold reheat.

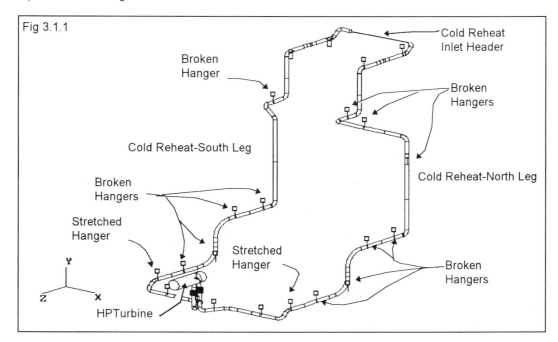

Fig 3.1.1

CAUSE OF DAMAGE

A waterhammer was the only logical source with enough force to result in so much damage. How would water get in a cold reheat steam system? During the time the repairs were being made, the boiler feed pumps were in operation for the boiler and supplying water to the steam attemperator valve which was at least leaking, if not open.

SOLUTION

We had to determine how to get the unit back on line as soon as possible. Because constant support hangers usually require ten to sixteen weeks for delivery, HEP had to determine if any other type hangers could be used. Saturday night was spent gathering data to prepare for an analysis. Sunday morning, a computer model was built of the system from and including the turbine up to the boiler connection. A standard analysis was run to determine which hangers were required for the piping to conform to ANSI B31.1 stress allowables. Various combinations of variable spring and rod hangers were tried, until a solution was settled upon (The new support scheme is shown in Fig 3.1.2). Sunday night, drawings (hand sketches) were prepared, and a bill of material was developed from the drawings. On Monday morning (which just happened to be a federal holiday), the bill of material was faxed to various vendors for delivery and price estimates. Fortunately, a hanger company had just opened a factory within a hundred miles of the plant and had all the components needed. An order was placed, and in the time it took our driver to go to the factory the order was assembled. On, Tuesday morning, the repairs were started and continued around the clock. Then Thursday morning, a fire was built in the unit, and it was back on line before noon.

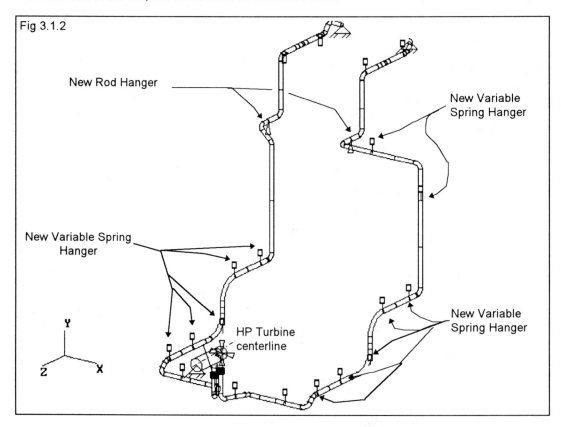

Fig 3.1.2

New Rod Hanger

New Variable Spring Hanger

New Variable Spring Hanger

New Variable Spring Hanger

HP Turbine centerline

MORAL

A proficient High Energy Piping team is good for a lot of megawatt hours.

CATEGORY 4

This section illustrates long term pipe movement (drift) in Main Steam and Hot Reheat piping systems in which the majority of the hangers are constant load. The Shawnee Fossil Plant requested technical assistance to investigate problems in the main steam and hot reheat piping on several of the nine sister units. The main steam piping had an abnormal deformation (S-shape) and repetitive cracks in the wye-block; whereas, the hot reheat had only repetitive cracking of the T-block.

EVENT #1
MAIN STEAM
INVESTIGATION

A system walkdown (both hot and cold) was conducted on unit 9 which showed the following (see sketch for main steam configuration Fig 4.1.1):

⇒ The west rigid support (8005) had drifted 2" west and was in contact with guiding steel, and the east support (8001) had drifted 1" east. Both rigid supports were off their slide centerlines, indicating a spread in the branch line.

⇒ Monitoring of the rigid supports (8001 & 8005) indicated only 4 inches of axial movement.

⇒ The constant support hangers at the top of the riser (8002 & 8006) were topped out and at the bottom of the riser (8003 & 8007) were bottomed out.

⇒ The main steam outlet header is not independently supported, but suspended by the crossover tubes and their supports. There are eight counterweight supports above the boiler similar to the old cotton bale scales with a 4-1 lever ratio. The outboard supports were almost "bottomed out".

⇒ A typical counterweight consists of a steel container (box) with a calculated amount of concrete and loose metal chips. The amount of chips varied from box to box as a result of flyash removal (vacuuming). Measurement of the counterweight contents revealed that the actual support load, varied from 8400 lbs to 9530 lbs. As subsequently discussed, the analysis identified the counterweight loads needed.

SOLUTION

A comprehensive piping model, including the boiler crossover tubes, was developed and a detailed analysis performed to evaluate the piping stresses and support loads for two conditions: As-Designed and As-Operating. The as-designed analysis provides optimum data to verify the original design data. The as-operating provides stress on piping and loads on supports and equipment due to the present condition of piping and supports.

The as-designed analysis results revealed the following:

1. The crossover tube counterweight support load is not identical for each of the eight supports as originally designed. The end supports should be larger due to the safety valves and the support spacing. The table below lists data for the as-designed, actual load, and the original load for each support.

2. The original total support load was very accurate when compared to the as-designed total load with load distribution being the exception.

3. When the actual support loads were calculated, they were 20,760 lbs less than required. This difference was caused when the counterweights were vacuumed to remove accumulated flyash.

4. The axial movement at the rigid supports was 8.5", but only 4" had been observed during the walkdown.

Main Steam Support Load Table

Support ID	Original Design Load (lbs)	Actual Support Capacity (lbs)	As-Designed Load (lbs)
8104 West	11,600	8830	14,000
8103	11,600	9530	12,000
8102	11,600	9100	10,400
8101	11,600	8400	9300
8111	11,600	9050	9400
8112	11,600	9530	10,600
8113	11,600	8700	12,400
8114 East	11,600	9000	14,800
Total Load	**92,800**	**72,140**	**92,900**

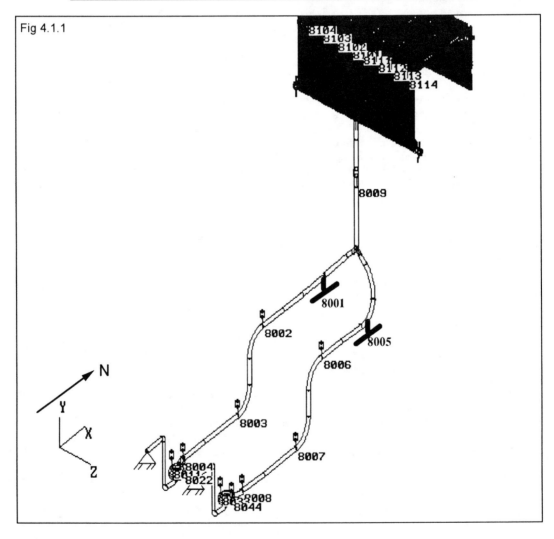

Fig 4.1.1

The as-operating analysis results, which include the actual support data, are as follows:
1. The piping stresses at the wye-block branch ends (weld locations) exceed the sustained code allowable by a factor of 2.0.
2. The wye-block sags 0.62 inches, and the branch lines move 0.4 inches outward at the rigid supports locations due to the deadweight loading.
3. Since the tube counterweight supports at the ends do not support the required load, the outlet header ends sag 0.85 (8104) and 1.35 (8114) inches respectively.

The analysis results confirmed that the counterweights supports have the following problems:
1. Their capacity is below both the as-designed and original loads.
2. The hangers should not have equal capacities. The end supports must carry larger loads because of the unequal spacing and safety valves on the ends of the header.

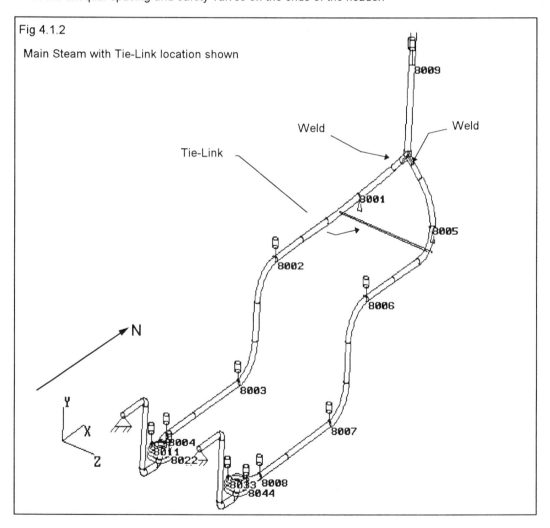

Fig 4.1.2

Main Steam with Tie-Link location shown

CONCLUSION/RECOMMENDATIONS

The actual loads are larger than the capacities of the installed counterweight supports. Therefore, the rigid supports were subjected to higher than design loads. The increased loads forced the branch lines to drift outward and caused the slides to bind. In addition, the outward movement of the branch lines imposed large bending moments and subsequently high stress in the wye block.

The effect of 42 years of thermal cycling, creep, and excessive load caused progressive drifting of the branch lines, which caused the rigid support guide to bind and resist the axial thermal growth. Consequently, the line was forced into the s-shape at the short riser in the system.

To alleviate the piping system problems, the following support modifications were made:
1. Put additional ballast in the counterweights to achieve the as-designed values.
2. Installed a tie link between the branch lines to ensure that the lines do not spread further and that the guides not bind. There is an expectation that the tie link will reverse the spread.
3. Adjusted the system hangers, so that they operated in their travel range.
4. Periodically, check ballast boxes for flyash accumulation and remove if deemed significantly heavy. Sheet metal covers should be considered to eliminate flyash collection in the ballast boxes.

EVENT #2
HIGH TEMPERATURE REHEAT

INVESTIGATION

A system walkdown (both hot and cold) was conducted which showed the following:

⇒ No major pipe distortion was observed.

⇒ The fourteen counterweights on some of the units had various amounts of weight as a result of flyash removal.

⇒ Additional weight had been placed in the counterweight buckets on Unit 2. Used train rails and pulverizer balls were utilized for supplementary ballast.

⇒ Measurement of the counterweight contents revealed that the actual support load varied from 5050 lbs to 5250 lbs.

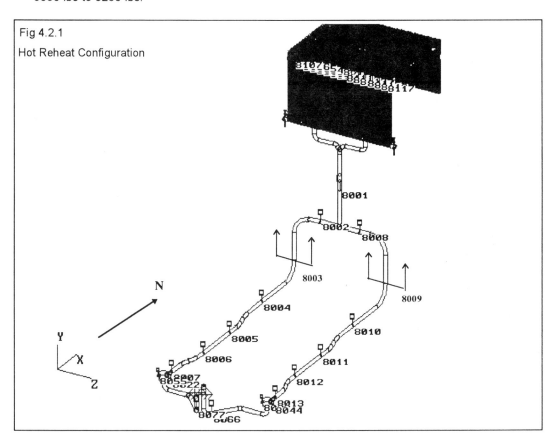

Fig 4.2.1

Hot Reheat Configuration

SOLUTION

A comprehensive piping model, (Fig 4.2.1 shows a general configuration) including the boiler crossover tubes, was developed and a detailed analysis performed to evaluate the piping stresses and support loads for two conditions: As-Designed and As-Operating. During model building for the analysis, the drawings revealed that the original crossover tubes (2" OD x 0.130" wall) had been replaced by heavier wall tubes (2 " OD x 0.148" wall). The thicker wall makes the tubes 13% heavier, and there was no evidence that the crossover counterweight had been modified to accommodate this extra weight.

The as-designed analysis results are as follows:
1. The load is not identical for each of the fourteen supports. The tabulation below lists data for the as-designed, actual load, and the original load for each support. Note that the end loads are larger due to the safety valves and the number of crossover tubes carried by each support.
2. The total original support loads (108,416 lbs) are 26% less than the total as-designed loads (136,590 lbs). The increased load can be attributed to the heavier tubes and the utilization of the computer analysis which provides more accurate load distribution of the entire piping system.
3. The actual total support capacity of 72,590 lbs was reduced to this value when the ballast boxes were vacuumed to remove accumulated flyash, which also removed the metal chips. As a result, the load difference (64,000 lbs) shifted to the two rigid supports at the base of the downcomer.

Hot Reheat Support Load Table

Support ID	Original Design Load (lbs)	Actual Support Capacity (lbs)	As-Designed Load (lbs)
8107 West	7744	5250	12,100
8106	7744	5200	10,500
8105	7744	5200	9560
8104	7744	5060	8900
8103	7744	5200	9020
8102	7744	5200	9020
8101	7744	5220	9200
8111	7744	5250	9150
8112	7744	5200	9010
8113	7744	5200	9020
8114	7744	5160	8900
8115	7744	5200	9560
8116	7744	5200	10,500
8117 East	7744	5050	12,150
Total Load	**108,416**	**72,590**	**136,590**

The as-operating analysis results, which includes the actual support data, are as follows:
1. The stress due to sustained loads at the T-block branch ends (weld locations) are 39,500 psi and 39,870 psi, which is 577% of the code allowable 6845 psi. A high bending moment due to deadweight loading results from unbalanced load distribution, i.e., deadweight shifting from fourteen crossover tube supports to the rigid vertical supports.
2. The T-block sags 2.67 inches due to deadweight.
3. Again, the rigid supports had to support the difference (64,000 lbs) in required load and actual load of the counterweight supports.
4. Periodically, check ballast boxes for flyash accumulation and remove if deemed significant.

CONCLUSION/RECOMMENDATIONS

The high temperature reheat crossover tube counterweight supports are significantly undersized. Fortunately, the rigid supports had been designed very conservatively and did carry the additional loads without damage. The shift in load caused the high stresses in the T-block.

The main recommendation was to add ballast to the counterweights to the as-designed values. A decision was made to use scrap train rail and steel shot for the additional ballast. With better load distribution, the anticipated result is elimination of T-block weld cracks.

CATEGORY 5

This section illustrates girth/saddle weld cracking in high temperature (main steam) piping due to inadequate flexibility or geometric configuration.

EVENT

On February 28, 1996, Tennessee Valley Authority (TVA) Paradise Fossil Plant unit 2 was forced to shutdown due to extensive weld cracks in the turbine bypass lines.
On June 28, 1996, similar findings of weld crack failures in unit 1 turbine bypass lines were identified. The weld cracks were at very similar locations on the turbine bypass piping of unit 2. Units 1 and 2 are sister units.

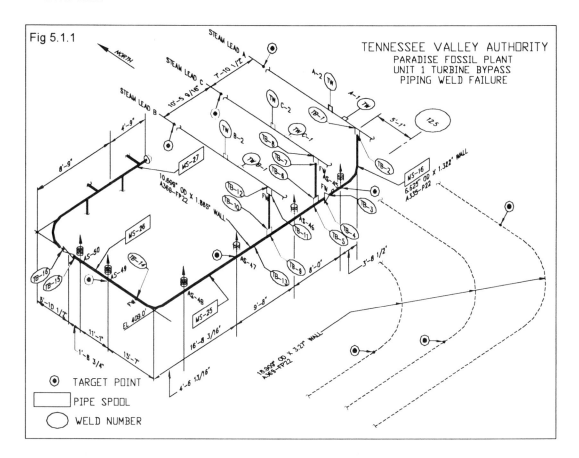

INVESTIGATION

Upon the discovery of the turbine bypass weld cracks in each unit, a nondestructive examination (NDE) was performed on thirteen girth and saddle welds (TB-1 to TB-13) utilizing the dry magnetic particle method. See Figure 5.1.1 for location of welds.

The weld examinations of both units revealed the following:

Weld Location	UNIT 1			UNIT 2	
	NDE Finding	Severity Rank (1)		NDE Finding	Severity Rank (1)
TB-1	11" indication @ 3:00 on top side of saddle weld.	II		8" indication @ East side, approx. 1/4" deep.	II
TB-2	6" indication @ 3:00 on top side of girth weld.	II		Indication @ East and West side, Approx. 1/4" deep.	II
TB-3	6" indication @ 3:00 to 9:00 on the left side of the weld. A 3.5" indication in the base metal @ 12:00 on the reducer between TB-3 and TB-4.	I		5" indication @ top, thru wall.	I
TB-4	No indication found.	III		No indication found.	III
TB-5	19" indication on the bottom side of the saddle weld.	II		Indication around circle of weld, approx. 3/4" deep.	II
TB-6	No indication found.	III		No indication found.	III
TB-7	No indication found.	III		Indication @ West side, approx. 1/8" deep.	II
TB-8	No indication found.	III		Indication @ West side, approx. 1/8" deep.	II
TB-9	28" indication on the bottom side of the saddle weld, and 2" indication @ 9:00 on the top of saddle weld.	II		Indication around circle of weld, approx. 3/4" deep.	II
TB-10	No indication found.	III		No indication found.	III
TB-11	3" indication on the top side of the girth weld, and a 5" indication the bottom side of the girth weld.	II		Indication @ North and South side, approx. 3/8" deep.	II
TB-12	No indication found.	III		No indication found.	III
TB-13	No indication found.	III		No indication found.	III

(1) I indicates severe or through wall crack, II indicates evident crack, and III indicates no crack.

Further investigation, in addition to the NDE results, support the following findings:
- The records indicate that these piping systems had not been previously inspected (both units have about 208,000 hours of operation), which is consistent with the practice to confine NDE to the large diameter main steam and hot reheat piping in the absence of known liabilities in the small diameter take-off piping.
- The worst failure location at TB-3 was in the fine-grained portion of the heat-affected zone of the base metal. This was established from boat samples taken at the through-wall crack on the small end of the reducer. By inference, it is expected that other cracks would show a similar morphology.
- Failures in the fine-grained portion of the heat-affected zone are associated with an overstressed condition in a direction perpendicular to the fusion line. For girth welds, the thermal expansion stresses are the most likely cause.
- A review of the design stresses (as-designed configuration) indicates that piping sustained and thermal expansion stresses (for all possible operating modes) are within the B31.1 code allowable limits, even though it was suspected that the "close coupling" of the short legs connecting the bypass lines to the main steam lines was an underlying cause.

ROOT CAUSE ANALYSIS

To demonstrate that the locations of the weld cracks are subjected to high thermal stress condition, a more representative analysis technique is utilized for unit 1 turbine bypass piping configuration. A different analysis technique is developed to simulate the actual piping system thermal growth in the vicinity of the subject welds area.

Since the actual piping thermal growth usually varies from the as-designed analysis based on the some of the following elements (facts):

- Inherent friction in the variable hangers and sliding supports which tends to increase as the plant gets older.
- Relative settlement of the main headers (such as the main steam lines) to which smaller pipe sizes are attached.
- Inadequate support design (short rods, bottomed-out hangers, inaccurate direction of horizontal travelers, etc.) which tends to change the movement in the supporting direction and restrict the movements in other directions.
- Plant physical obstruction and interferences which impact the pipe movements.

Therefore, to take into account the actual piping thermal growth to be represented in the as-operating stress evaluation, several locations (targets) around the subject welds area were monitored during cold and hot conditions. The monitoring of the target locations, in form of location measurements and hanger movements, enabled us to obtain movement data that represented the actual pipe movements for the turbine bypass lines under consideration.

The piping analysis evaluation was then performed using actual thermal movements forcing piping displacements. This evaluation produced representative and more realistic thermal expansion stress values at the subject weld locations. The results of this evaluation may be summarized as follows:

- The turbine bypass lines are subject to high thermal stresses which exceed the code allowable limits at eleven (out of thirteen) weld locations, while the sustained stresses remain within the code allowables.
- The highest thermal stress was found at the small end of the reducer (TB-3). This agrees with the NDE findings as of the worst crack indication.
- In general, there is a correlation between the thermal stress value and the severity of the weld crack, with the exception of two weld locations (TB-1 and TB-7).

Weld Location	Piping Thermal Expansion Stress (psi)	Allowable Thermal Expansion Stress (psi)	Severity Rank of Weld Crack
TB-3	81,514	20,176	I
TB-11	52,133	20,176	II
TB-7	48,858	20,176	III
TB-5	43,118	20,176	II
TB-9	35,756	20,176	II
TB-2	35,568	20,176	II
TB-8	31,616	20,176	III
TB-4	30,786	20,176	III
TB-12	30,550	20,176	III
TB-6	26,730	20,176	III
TB-1	23,735	20,176	II
TB-13	18,002	20,176	III
TB-10	17,938	20,176	III

CONCLUSION/RECOMMENDATIONS

Based on the above evaluation results, it is evident that the main steam piping actual (operating) thermal growth induces extremely high thermal stresses into the turbine bypass lines. The main contributing factor to this condition is the lack of piping flexibility to accommodate the actual differential thermal growth between the main steam piping and the turbine bypass lines. The lack of flexibility, in this case, could have not been detected from the original as-designed analysis.

It was recommended to repair the welds immediately and attention be given to the system hangers settings, as it was observed that three main steam supports near the turbine bypass lines were bottomed out during the hot condition. As a result, the system hangers should be checked and adjusted to ensure they are functioning properly and their movements are within the hanger working range for hot and cold conditions. Also, the entire piping system should be inspected to identify any potential thermal bindings, obstructions, or interferences that could hamper or divert the piping thermal growth. In addition, a proposed new routing for the turbine bypass lines to provide more flexibility and reduce the thermal expansion stress was also recommended for both units 1 and 2.

The piping flexibility analysis, for design purposes, is mainly used to establish compliance with the piping code stress limits, and calculate support loads. However, actual piping thermal growth, especially for high temperature lines, may significantly differ from the idealistic piping model condition used in the as-designed analysis condition, as previously mentioned. Furthermore, the as-designed analysis is used as a design tool based mainly on the theory of elasticity. As such, it does not phenomenologically model creep and stress relaxation which can play major roles in actual pipe failures.

When there are actual weld failures, as in the turbine bypass piping, we are in a more enlightened position to search for answers than in the general case. We often find that the "as-designed" stresses are not exceptionally high with respect to the B31.1 code allowable stress limits, so the piping flexibility analysis does not have much success at predicting future weld failures. Additionally, weld failures sometimes result from "metallurgical complexity" that isn't formally considered in the piping flexibility analysis. However, the analytical approach, presented here, of simulating the actual piping thermal growth based on field measured pipe/hanger movement data provides more realistic (localized) stress values which may be the answer or a contributing factor to the cause of weld failures. This analytical approach can also be utilized for highly sensitive geometric configuration (limited flexibility) piping areas to predict potential weld failures before they occur.

SYNOPSIS

The computer software for static and dynamic piping analyses is a useful tool that can be used to determine the root causes and solutions to a variety of operational problems related to piping systems. As this paper illustrated, the use of the computer-based analysis in a systematic way does not necessarily yield results that support or conform to actual piping performance and emerged problems. However, with thorough field examination of all the elements involved in an identified problem, combined with a computer model that captures and/or simulates the examined field condition, then the analysis results will not only confirm the problem event, but it will also provide a path to a solution.

Although the computer analysis is an essential tool in solving piping related problems, a sound engineering judgment, based on careful examination of the aspects involved in any piping related problem, is still fundamental to turn the computer-based analysis into a useful technique. It is essential to recognize that the success of resolving any of these operational problems can only be measured by monitoring the involved components to ensure that the problem event has become nonrepetitive.

ACKNOWLEDGEMENTS

The authors wish to express their sincere gratitude to the following:
- **Blaine W. Roberts** of TVA Fossil and Hydro Power for his philosophical suggestions, moral support during the course of this paper development, and his daily support as a supervisor.
- **Larry R. Spiva** for being a part of the High Energy Piping Team which made the category 3 event a major success at Kingston Fossil Plant.
- **Robert S. Love**, a member of the High Energy Piping Team, who provides the field knowledge and experience to insure that recommendations can be and are installed correctly.
- **David H. Hardgrave**, a member of the High Energy Piping Team, who provided some of the drawings for this paper and all of the drawings required in our daily work.

REFERENCES

1. ASME/ANSI B31.1 Power Piping Code for Pressure Piping
2. CAESAR II - Pipe Stress Analysis program - Engineering software from COADE, INC.
3. AUTOPIPE - Pipe Stress Analysis program - Engineering software from Rebis Industrial Workgroup Software.

****

PVP-Vol. 356, Integrity of Structures and Fluid Systems,
Piping and Pipe Supports, and Pumps and Valves
ASME 1997

QUANTIFICATION OF NONLINEAR SOIL RESTRAINT
IN THE ANALYSIS OF RESTRAINED UNDERGROUND PIPING

by

Evans C. Goodling, Jr.
Parsons Power Group Inc.
Reading, PA

ABSTRACT

Analysis of restrained underground piping systems
can be easily performed with the use of a personal
computer-based general purpose pipe stress program.
However, modeling of the restraining effects of the
soil at the pipe/soil interface presents some uncertain-
ties to the analyst due to the non-linear effects of soil
resistance to lateral movement of the pipe. This paper offers some guidance in developing the soil spring
rates to be used in the computer model. Quantifica-
tion of the spring rates is based on experimental results from actual tests performed and described by
Audibert et al. Approximate effects of friction are
also addressed, in order to enable an analyst to approximate the total soil environment in the piping
analysis. The primary objective of the paper is to assist the piping engineer in analyzing buried piping in
accordance with the guidelines of Appendix VII of the
ASME B31.1 Code for Power Piping.

INTRODUCTION

Initially, procedures for analyzing buried piping were
driven by the nuclear industry, and were primarily
aimed at qualifying safety class piping for the effects
of seismic soil strain and thermal expansion. However, guidelines for analyzing the stresses in restrained
underground piping have yet to be incorporated in the
nuclear piping Code, Section III of the ASME Boiler
and Pressure Vessel Code. Until recently, thermal
expansion stresses on restrained underground piping
were covered by specific equations only in the ASME
B31.4 and B31.8 Codes for oil and gas pipelines [1][2].
Peng [3] published guidelines for calculating stresses,
including approximate equations for estimating passive soil bearing and friction reactions. Audibert and
Nyman [4][5] and Trautmann and O'Rourke [6] defined the non-linear relationship between force per
unit length and horizontal displacement of piping buried under typical sand backfill. In 1992, Nonmanda-
tory Appendix VII of the B31.1 Code for Power Piping
[7] presented some suggested procedures for qualification of restrained underground piping, using linear
soils springs to simulate passive soil resistance to lateral displacement.

The paper is intended to supplement the B31.1 Code
through commentary and extension of the part of Appendix VII dealing with the passive behavior of the
soil at the pipe surface. It shows the analyst how to
readily estimate the number and magnitude of soil
springs required for the pipe stress computer model of
a pipeline buried and backfilled with sand. It also offers some suggestions on how to incorporate Peng's
treatment of axial friction effects into the stress model. The paper touches only lightly on the geotechnical
aspects of soil and its resistance to lateral displacement of buried piping. For a more comprehensive
treatment of soil characteristics, the reader is referred
to [5], [6], and [9].

DEFINITIONS AND NOMENCLATURE

Virtual anchor - the point on a pipe run where external passive bearing and friction forces along the pipe axis equal the force required to completely restrain thermal expansion, resulting in zero relative movement between the pipe and the soil [7].

Effective length L' - the length of a pipe run from the virtual anchor to a change in direction; sometimes called the active length, because it is within this length that relative displacement of a thermally-expanding pipe occurs with respect to the confining soil.

Axial length L_1 - total length a pipe run containing a virtual anchor.

Transverse run length L_2 - the length of a pipe run on which soil lateral bearing forces exist.

Influence length L_i - that portion of a transverse pipe run that becomes laterally displaced due to thermal expansion of an adjacent connected axial run; the length of pipe in the computer model that is divided into elemental lengths with soil springs imposed [7].

Modulus of subgrade reaction k -the rate of soil resistance to lateral displacement of a pipe [7]. The values for k that are calculated by the procedures outlined in [9] differ somewhat depending on whether the lateral directions are horizontal or vertical. Conservatively, the resistance to upward movement (contraction of a vertical pipe run) may be considered the same as for a horizontal pipe run with the soil weight effects added. Resistance to downard movement may be considered as rigid for most expansion stress analysis [7].

System characteristic ß - a function of soil modulus of subgrade reaction k, pipe section moment of inertia I, and pipe modulus of elasticity E used to predict the behavior of pipe on a continuous elastic foundation [7].

ANALYSIS

Linear soil springs

Calculation of the soil springs $k_{i,j}$ to be imposed at the center of each elemental length along the influence length of a transverse pipe run is covered in B31.1 Subsection VII-4.2.2 [7] by the equation -

$$k_{i,j} = kdL, \text{ lb/inch (N/mm)} \tag{1}$$

where $k_{i,j}$ are the vertical and lateral spring rates imposed at the center of an element of length dL. The variable k is the modulus of subgrade reaction calculated in accordance with Subsection VII-3.2.2 [7]. For the horizontal direction, $k = k_h$ as calculated by an equation proposed originally by Trautmann and O'Rourke [6] -

$$k_h = C_k N_h wD, \text{ N/mm}^2 \text{ (lb/in}^2\text{)} \tag{2}$$

where C_k = dimensionless horizontal stiffness factor for backfill (use 20 for loose sand, 30 for medium backfill, or 80 for dense or compacted backfill [6]).

N_h = dimensionless horizontal force factor
$\approx 0.285H/D + 4.3$ for a typical soil internal friction angle of 30°.

w = soil density, weight per unit volume

D = pipe outside diameter inches (mm)

Equation (2) assumes linear elastic behavior of the soil, with the calculated value of k_h independent of the magnitude of horizontal displacement of the pipe against the soil.

The Audibert-Nyman approach

A more rigorous approach to determining soil spring rates is built on the results of actual tests reported by Audibert and Nyman [4][5] and later confirmed by Trautmann and O'Rourke [6]. In order to recognize the hyperbolic relationship between displacement and passive restraining force, Reference [9] proposes the following equation:

$$p = y/(A'+B'y) \tag{3}$$

where p = unit passive force in compression, lb/in (N/mm)

y = horizontal displacement, inch (mm)

$A' = 0.15 y_u/p_u$

$B' = 0.85/p_u$

$p_u = wHN_{qh}D$, lb/inch (N/mm)

$y_u = 0.02$ to 0.03 H+D/2 for dense sand

w = soil density, weight per unit volume

H = depth below grade to pipe centerline, inches (mm)

N_{qh} = horizontal bearing capacity factor

The fundamental variables are further defined as follows:

The passive unit force p, the dependent variable in (3), is the reaction force of the soil per unit length of the pipe in response to an imposed lateral displacement y.

The ultimate passive unit force p_u is the maximum resistance capability of the soil to lateral displacement of a pipe.

The ultimate displacement y_u is the value of displacement at which the ultimate passive force p_u is reached. It is the point at which further displacement of the pipe will not result in any increase in passive resisting force. The horizontal bearing capacity factor N_{qh} is a dimensionless factor which is a function of pipe depth H below grade, pipe diameter D, and internal soil friction angle Φ. Curves of N_{qh} as a function of H/D are plotted in Fig. 2 for various values of Φ.

The nonlinear soil spring rate k_h in force/unit displacement /unit length is expressed by the following equation:

$$k_h = p/y, \text{ lb/inch}^2 \text{ (N/mm}^2\text{) (4)}$$

Then, the lateral soil spring rate for an element of length dL is expressed by -

$$k_{i,j} = dLp/y, \text{ lb/inch (N/m)} \quad (5)$$

Calculation procedure

The step-by-step calculation of soil springs for the buried pipe stress computer model using the Audibert-Nyman approach [5] is described as follows:

1. Tabulate values for p as a function of y and plot using Equation (3) for the diameter D of the pipe to be analyzed.

2. Calculate the equivalent net friction force F_f in each run of pipe, using the methods defined in [7]. The equivalent net friction force F_f can be imposed axially near the bends in order to simulate the effect of friction-induced axial strain, which will partly offset the displacement of the transverse runs.

3. Determine the number n and length dL of elements in the influence length L_i for each transverse run that will cause soil displacement.

4. Note that $ndL \geq L_i$. Develop the computer model for the piping system, with a node at the center of each element for which a transverse soil spring rate is to be calculated for incorporation into the model.

5. Make an initial estimate of lateral displacement of each soil spring node in the buried piping system. A good set of initials values would be the theoretical unrestrained thermal displacements of the piping system.

6. For each of the soil spring node points, enter the table or curve developed in Step 1 with an initial estimated value for lateral displacement y from Step 5. in order to determine the corresponding unit restraining force p of the soil. Calculate a value for spring rate k for each spring rate node by the equation $k = dL(p/y)$ and input these values in the pipe stress computer model.

7. Analyze this model of the buried piping system with the appropriate net equivalent friction forces imposed axially at each elbow and with the soil spring k values imposed on each spring rate element. From the operating case (sustained + thermal expansion) of the computer, determine the calculated transverse movement at each soil spring node point and compare it with the movement used for setting the value of soil spring rate in Step 6. If there is a significant difference in transverse movement at any of the elements, repeat Step 6 to develop new spring rates based on the calculated movements of the elements, and then repeat Step 7 to reanalyze the model. Repeat this step until convergence is reached. Two or three iterations will produce satisfactory convergence.

8. After convergence is reached, the final computer run is then be used to qualify the buried piping system to the appropriate ASME Code for pressure piping.

Note that for small values of y, the curve of p versus y may be approximated by a straight line, in which region the values for spring rate k calculated by $dL(p/y)$ will be essentially constant.

As the analyst develops a feel for accurately estimating the initial displacement of the transverse runs, convergence may be reached more quickly, perhaps in just two runs. Similarly, for relatively low magnitude of pipe thermal expansion, the soil spring rate will vary little and may be considered constant.

SAMPLE PROBLEM

As an example, the sample problem from Section VII-6.0 of B31.1 Appendix VII [7] is considered here, with soil springs and pipe stresses calculated in accordance with the 8-step procedure presented above and using an element length dL of 3x nominal pipe diameter.

1. Values for p vs. y are calculated from Equation (3), tabulated, and then plotted as follows (see Figure 3):

$$p = y/(A'+B'y) \qquad (3)$$

$$A' = 0.15y_u/p_u$$

$$y_u = 0.025(H+D/2)$$
$$H = 144.00 \text{ inches}$$
$$D = 12.75 \text{ inches}$$
$$y_u = 3.76 \text{ inches}$$

$$p_u = wHn_{qh}D$$
$$w = 130.00 \text{ lb/cf}$$
$$H/D = 11.29$$
$$n_{qh} = 7.00$$
(extrapolated from Figure 2)
$$p_u = 966.88 \text{ lb/inch}$$

$$A' = 0.000583 \text{ in}^2/\text{lb}$$
$$B' = 0.85/p_u$$
$$= 0.000879 \text{ in/lb}$$

See Table 1 for calculated values of p as a function of y.

2. The net equivalent friction forces F_f are calculated in accordance with Equation (13) of [7], and are found to be:

   a. At Node 85,
      $F_f = F_z$ = -76,605 lb (-340,739 N)

   b. At Node 115,
      $F_f = F_x$ = -22,410 lb (-99,680 N)

   c. At Node 295,
      $F_f = F_x$ = 22,410 lb (99,980 N)

d. At Node 325,
   $F_f = F_z$ = 8,964 lb (39,872 N)

These values are a function of the integration of confining pressure along the slippage length and an estimated coefficient of friction of 0.3 at the pipe/soil interface.

Table 1

p as a function of y (from Step 1)

y inches	p lb/inch
0.10	149.00
0.20	263.49
0.30	354.21
0.40	427.87
0.50	488.86
0.60	540.20
0.70	584.01
0.80	621.83
0.90	654.81
1.00	683.83
1.10	709.56
1.20	732.52
1.30	753.15
1.40	771.78
1.50	788.68
2.00	854.17
2.50	898.95
3.00	931.51
3.50	956.25
3.60	960.50
3.70	964.55

SI conversion:
   1 inch = 25.4 mm
   1 lb/inch = 0.175 N/mm

3. The influence length L_i is found from Section VII-6.3.1 of [7] to be 202 inches (5.131 m). The unrestrained thermal expansion displacements dx and dz at the bends at Nodes 100 and 310 are as follows: at Node 100, dx = 0.265 (6.731 mm) inches and dz = 0.906 inches (23.01 mm) at Node 310, dx = -0.265 inches (-6.731 mm) and dz = -0.106 inches (2.692 mm). These are based on the thermal expansion of the piping at 140°F (60°C), assuming virtual anchors at Nodes 1 and 200 and a real anchor at Node 420, but

without any restraint due to passive soil resistance, either by friction or transverse bearing.

The length of the soil spring elements are typically set at 2 or 3 times the nominal piping diameter. For this example, an element length of approximately 3D, or 36 inches (0.914 m), is chosen. The number of elements n is then found from Section VII-6.4.2 of [7] to be 6. The center of each element is given a node number in the computer model.

4. The computer model is then constructed with the soil spring nodes located at the center of each element within the influence length approaching and departing each bend node.

For example, in this problem, there are 6 soil spring nodes on each side of each of the two bends, as follows: approaching bend Node 100 are soil spring nodes 10, 25, 40, 55, 70, and 85, each located 36 inches (0.914 m) apart and Node 85 located 18 inches (0.457 m) from the bend intersection; departing the bend are soil spring nodes 115, 130, 145, 160, 175, and 190, spaced in a manner similar to the nodes approaching the bend.

Likewise, soil spring nodes are established at the approach to and departure from the other bend, Node 310.

5. The first step in estimating the lateral displacements within the influence lengths adjacent to each bend is to calculate the thermal displacement at each bend, considering the pipe to be unrestrained laterally except at the virtual and actual anchors at nodes cited in Step 3 above. Assume initially that the lateral displacement of each of the soil spring nodes within the influence length is inversely proportional to its distance from the bend intersection. Accordingly, the displacement is maximum at the bends and decreases as the limit of the influence length is approached.

For this example, values for the first computer iteration are estimated as given in Table 2 (see Figure 4 for the location of the soil spring element centroidal nodes).

Table 2 - Initial displacements

Adjacent to bend at Node 100

Node	Initial lateral displacements (inch)
10	x=0.005
25	x=0.052
40	x=0.100
55	x=0.147
70	x=0.194
80	x=0.241
100	x=0.265, z=0.906
115	z=0.825
130	z=0.664
145	z=0.502
160	z=0.341
175	z=0.179
190	z=0.018

Adjacent to bend at Node 310

Node	Initial lateral displacements (inch)
220	z=-0.002
235	z=-0.021
250	z=-0.040
265	z=-0.059
280	z=-0.078
295	z=-0.097
310	x=-0.265, z=-0.106
325	x=-0.241
340	x=-0.194
355	x=-0.147
370	x=-0.100
385	x=-0.052
400	x=-0.005

6. The curve or table developed in Step 1 is used to determine the initial unit soil restraining force p for each value of lateral displacement x or z from Step 5 above. The y term in Equation (3), from which the table and curve are developed, is a general expression for transverse displacement, and represents the specific x and z values for lateral displacement in this problem. These

111

values for p are then used to calculate the soil spring rates $k_{x,z}$ from the equations $k_x = dL(p/x)$ or $k_z = dL(p/z)$, depending on the direction of lateral movement. For the first iteration, the spring rates given in Table 3 are incorporated in the model.

Table 3

Spring rates based on the initial displacements of Table 2

Node	k_x, lb/inch	Node	k_z, lb/inch
10	61,000	115	27,500
25	57,000	130	30,800
40	54,000	145	35,000
55	50,500	160	40,900
70	47,500	175	48,500
80	45,000	190	60,000
325	45,000	220	61,000
340	47,500	235	60,000
355	50,500	250	58,200
370	54,000	265	56,700
385	57,000	280	55,000
400	61,000	295	54,000

7. These spring rates are incorporated into the model, which is then executed to determine theoretical displacements for comparison with those used to calculate initial values for the lateral soil springs. These new values for displacements are then used to recalculate the soil springs, which are used iteratively in the model until there is convergence, i.e. no further change in displacements.

The lateral displacements and spring rates at convergence (after 3 iterations for this example) are given in Table 4.

8. The maximum B31.1 Code stresses for both the operating case (sustained + expansion) and the expansion case occur at Node 85 for all iterations. Table 5 lists the resulting Case 1 (operating) stresses and axial reactions for both the linear soil spring example based on Equation (1) and the nonlinear spring example based on Equation (5).

Table 4

Lateral displacements and soil spring rates at convergence (Compare these values with the initial values for displacements in Table 2 and initial soil spring rates in Table 3.)

Node	Lateral displacement (inch)	soil spring rate k_x, k_z (lb/inch)
10	x=-0.021	60,000
25	x=-0.009	61,000
40	x=0.011	60,700
55	x=0.049	57,500
70	x=0.109	53,000
80	x=0.181	48,400
115	z=0.363	39,900
130	z=0.195	47,700
145	z=0.074	55,400
160	z=0.007	61,000
175	z=-0.024	59,500
190	z=-0.038	58,400
220	z=0.010	61,000
235	z=0.005	61,000
250	z=-0.006	61,000
265	z=-0.028	59,600
280	z=-0.064	56,300
295	z=-0.107	53,100
325	x=-0.215	46,000
340	x=-0.111	52,900
355	x=-0.039	58,300
370	x=-0.004	61,000
385	x=0.007	61,000
400	x=0.004	61,000

DISCUSSION OF RESULTS

Refer to Table 5 for a summary of the analysis results. The value for the linear lateral soil spring case based on [6] and [7] for this example problem is 22,700 lb/inch (3975 N/mm). Using this value at the center of each 36-inch (914 mm) long element in the analysis model (which also includes an equivalent passive axial friction force) results in elbow stresses of 11,850 psi (81,700 kPa) at Node 85 and 7,114 psi (49,040 kPa) at Node 295. The axial load on the anchor at Node 420 is 13,997 lb (62,260 N).

The values for lateral soil springs on the elements based on the nonlinear test results reported in [4] and [5] range from a low of 39,900 lb/inch (6988 N/mm) at Node 115 (the point of maximum lateral displacement) to a high of 61,000 lb/inch (10,683 N/mm) for all locations where lateral displacement is small. Since this example is a 2-dimensional problem, vertical support is considered fixed at all lateral soil spring nodes. Using values for lateral soil spring based on convergence values of lateral displacement (and including the same equivalent axial friction force as for the linear case), results in elbow stresses of 12,169 psi (83,900 kPa) at Node 85 and 8,135 psi (56,090 kPa) at Node 295. The axial load on the anchor at Node 420 is 19,366 lb (86,140 N).

It appears therefore that the ratio of operating stresses for the non-linear case vs. the linear case is 12,169/11,850 or less that 1.03 at the bend at Node 85 (the point of maximum Code stress) and 8,135/7,114 or 1.14 at the other bend, at Node 295. The ratio of axial loading on the Node 420 anchor is 19,366/13,997 or 1.38. The ratio of axial loadings at the virtual anchor Node 1 is about 1.10. This suggests that where the piping runs are long and where axial friction at the pipe/soil interface has the opportunity to develop completely, elbow stresses and anchor loads do not appear to vary much between the cases where lateral soil restraint effects are treated as linear or non-linear. However for relatively short runs, where friction force is a smaller proportion of the net forces in the pipe, the theoretical operating stresses and anchor loads are higher for the non-linear lateral soil spring case.

CONCLUSIONS AND RECOMMENDATIONS

This paper started off with the thesis that a piping analysis would yield significantly higher stresses and anchor loads if the soil spring restraints in the analysis model were based on the nonlinear results obtained in tests conducted by Audibert and Nyman. However it appears that the differences are not very significant in long runs where soil friction along the pipe axis appears to play a more important role than lateral displacement of the transverse runs. However, in shorter runs, where passive axial friction does not develop fully, the increase in theoretical stresses and anchor loads is more significant.

It must be noted that this paper is based only on a single typical pipe run, the same example given in [7].

Therefore, the reader is advised to perform further trials with other piping configurations in order to judge the value of this approach in the analysis of re-strained buried piping, and to decide whether the presumed greater accuracy of the results is worth the extra effort. In lieu of the manual iteration used for the example in this paper, the reader might choose to perform a nonlinear analysis using commercially available software. Although it is relatively expensive, soil structure interaction analysis may also be used if greater accuracy of results is sought. However, the uncertainty of soil properties such as modulus of subgrade reaction and friction at the pipe/soil interface would make a more rigorous (and expensive) analysis difficult to justify for most applications.

ACKNOWLEDGMENTS

The author gratefully acknowledges the cooperation of the Parsons Power Group Inc. for the computer time required to perform the various analysis iterations. The helpful review comments by the author's colleagues, particularly those by James Yarger, Christopher Rickert, and Dr. Samir Serhan are greatly appreciated.

Table 5

Comparison of stresses and anchor loads
(SI conversion: 1000 psi = 6985 kPa; 1000 lb = 4.448 kN)

	Operating case stresses, psi		Anchor loads, lb (axial)	
	Node 85	Node 295	Node 1	Node 420
Linear $k_{x,z}$	11,850	7,114	91,640	13,997
Non-linear $k_{x,z}$				
Initial run:	10,476	7,524	95,972	19,130
Converged run:	12,169	8,135	100,658	19,366

REFERENCES

1. Goodling, E.C., "Buried Piping and the ASME Codes," Fifth International Symposium of Freight Pipelines, Philadelphia., October, 1985, Journal of Pipelines, Vol. 6 (1987), pp. 117-120.

2. ——— "Transportation of Hazardous Liquids by Pipeline," Part 195, Title 49, Code of Federal Regulations, revised as of October 1, 1994.

3. Peng, L.C., "Stress Analysis Methods for Underground Pipe Lines, Parts 1 and 2" Pipeline Industry, April 1978 and May 1978.

4. Audibert, J.M.E., and Nyman, K.J., "Coefficients of Subgrade Reaction for the Design of Buried Piping," Proceedings, Second ASCE Specialty Conference on Structural Design of Nuclear Power Plant Facilities, New Orleans, 1975, Vol. 1A, pp. 109-141.

5. Audibert, J.M.E., and Nyman, K.J., "Soil Restraint Against Horizontal Motion of Pipe," Journal of the Geotechnical Engineering Division, American Society of Civil Engineers, Vol. 103, No. GT10, October 1977, pp. 1119-1142.

6. Trautmann, C.H., and O'Rourke, T.D., "Lateral Force-Displacement Response of Buried Pipes," Journal of the Geotechnical Engineering Division, American Society of Civil Engineers, Vol. 111, No. 9, September 1985, pp. 1077-1092.

7. Appendix VII of ASME B31.1b-1996 Code for Power Piping, American Society of Mechanical Engineers, New York

8. Goodling, E.C. "Restrained Underground Piping - Some Practical Aspects of Analysis and Design," Third U.S. Conference on Lifeline Earthquake Engineering, American Society of Civil Engineers, Los Angeles, August 22-24, 1991.

9. Nyman, D.J., et al, Guidelines for the Seismic Design of Oil and Gas Piping Systems, Committee on Gas and Liquid Fuel Lifelines of the American Society of Civil Engineers Technical Council on Lifeline Earthquake Engineering, New York, 1984.

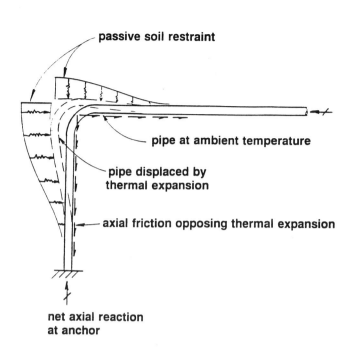

Figure 1 - Passive compressive soil reaction to thermal expansion displacement of a buried pipe bend

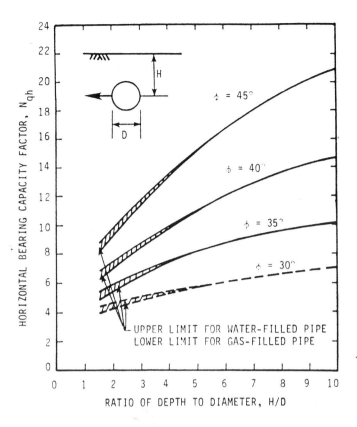

Figure 2 - Horizontal bearing capacity factor Nqh as a function of depth to diameter ratio of buried pipelines [6][9]

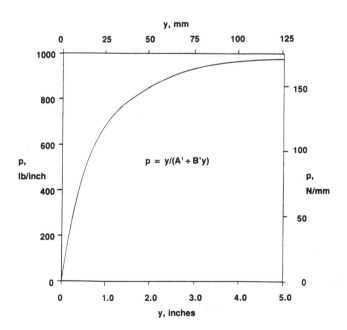

$$p = y/(A' + B'y)$$

Figure 3 - Soil reaction p vs. lateral displacement y for a buried pipe

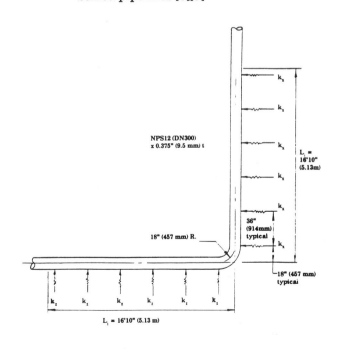

Figure 4 - Spacing of lateral soil springs over the influence length Li [7]

115

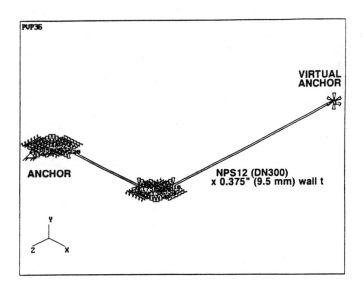

Figure 5a - Isometric plot of example problem from
B31.1 Appendix VII [7]

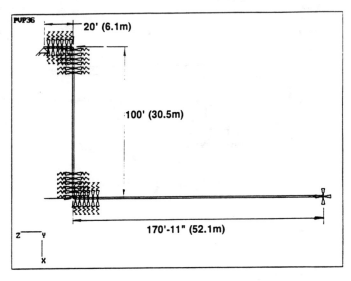

Figure 5b - Plan view plot of example problem from
B31.1 Appendix VII [7]

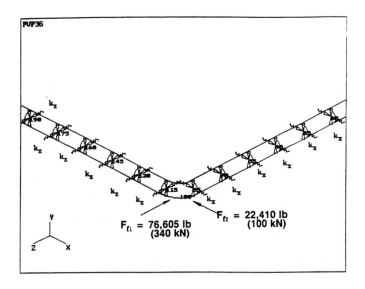

Figure 5c - Soil spring nodes and equivalent friction
reactions at bend at Node 100 [7]

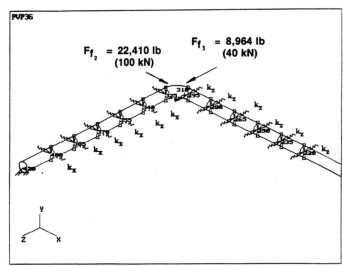

Figure 5d - Soil spring nodes and equivalent friction
reactions at bend at Node 310 [7]

PVP-Vol. 356, Integrity of Structures and Fluid Systems,
Piping and Pipe Supports, and Pumps and Valves
ASME 1997

MOMENT RESTRAINT SUPPORT FOR 84 INCH COMPRESSOR SUCTION LINE

Robert A. Robleto   Brown & Root Inc. Houston Texas

Abstract

In theory a device which constrains moments
without interfering with required translations is an
innovative solution to controlling forces and moments on load sensitive equipment. Theses fabled
devices have been mentioned in the literature [1] but
rarely seen in the field. This paper describes the extensive design confirmation of such a device. The
analysis includes both beam analysis pipe stress program and thin shell finite element analysis.

The analysis revealed the sensitivity of structural stiffness to the successful use of the device. In
addition, local stresses in the large OD pipe were
considered. In addition the finite element analysis
revealed significant differences in both the stiffness
and stress intensification factors of the large OD
pipe where D/t exceeded 150.

The design of the Moment Restraint Device was
unique, because the constraints in the field. Underground pipes prevented the most direct approach to
the design. The design reveals both the robustness of
the design concept but also the ingenuity of the support designer. The design concept can save on unnec-
essary piping loops delivering required suction head
on pump and compressor inlet piping while maintaining low forces and moments on equipment nozzles.

INTRODUCTION

The design of the pipe suction from the
ethylene scrubber to the suction of the ethylene compressor is constrained to meet the
requirements of NEMA 23 for external loads
applied to the flanges of the pipe connections
on the compressor. The piping system starting at the compressor has two 42 inch lines.
These lines are expanded to 60 inch and
then combined at the 84 inch manifold.
There a single 84 inch line leads to the
scrubber. The scrubber is a large horizontal
vessel, the piping to the suction of the compressor comes off the top of the vessel. The
piping system is shown from a plot from the
pipe stress model in Fig. 1.

Due to steam out conditions, it is possible for the pipe to experience full vacuum.
The hazard review requires that a temperature of 85° C (185° F) which considers failure of circulation pump and backup be used
as the design temperature. These conditions
are not coincidental but are considered in
combination with other temperatures and
pressures. This condition presented a design

117

temperature beyond temperatures considered at similar plants where solar heating of the pipe constitutes the design temperature on the suction piping. Meeting the requirements of NEMA 23 [5] is difficult for average size piping with one expansion in size, but meeting the requirements for 84 inch pipe presents an extraordinary problem. One that requires an ex-

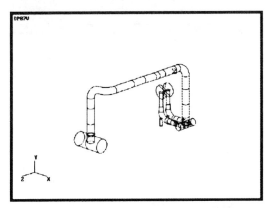

Figure 1: pipe stress beam model

BACKGROUND

A moment restraint system or MRS constrains one or more rotational movements at a point on a piping system while allowing one or more translation movements at the same point. The constraint can be described as a restraint within a constraint. The internal constraint prevents rotation, however the constraint itself can be guided in two other planes allowing translation. A similar orthogonal constraint can completely control all six degrees of freedom of a rigid body. The six degrees of freedom are three orthogonal rotations and three orthogonal rotations.

Theoretically all combinations of constraint such as two rotations constrained and all translations allowed with one rotation allowed. A simple example such as shown in Fig. 2 may be a rectangular slot in which a square tube is inserted perpendicular to the plane of the slot. The other end may be welded to a

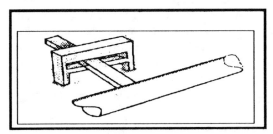

Figure 2: simplistic moment restraint

rigid body that is being constrained. The rigid body may move in and out of the plane of the slot, side to side in the slot, but in this case it is constrained in the vertical. The rigid body is also constrained in the in-plane rotational direction. And one out of plane rotational direction. The horizontal out of plane rotation is free.

BEAM ANALYSIS

The use of a beam analysis model for large OD pipe with D/t greater than 100 is inaccurate. Local bending and subsequent coupling makes it difficult to predict the resultant forces and moments due to fixed displacements caused by thermal growth. The following is a study of nonconcurring displacements for an "L" shaped structure. The structure is much more flexible than a beam analysis would predict.

Beam vs. Shell Element Model Description:

A beam analysis program was used to develop the forces and moments. The advantage of a beam analysis is that iterations are fast. It also gives better results at the boundaries than a finite element shell model. A shell model requires a spoked wheel structure to sum the loads at the center point of a pipe.

A study was performed to compare a beam model with a finite element shell model. A straight, bend, straight pipe model was created and individual loads were applied to a free end while the other end was held fixed. Runs were made with a single load in a unique degree of freedom were applied. Where no symmetry existed between positive application of the load and negative application of the load, two runs were made, one in the positive direction and one in the negative direction. The following describes the geometry and loading of the model.:
- 84 inch OD pipe D/t=168
- B31.3 [2] Appendix D compared with finite element model.
- Fifty foot rise, elbow, 26 ft 6 inch horizontal in X direction
- Anchor at base of vertical, unit displacements applied at end of horizontal. 55 psig internal pressure

- Loads of 1000 lb. and 1,000,000 inch pounds are applied singly in each direction. Only the motion in the direction of the load is free to move at the free end of the pipe.

From the results of the study shown in the table in appendix A the following is indicated:

- In general finite element shells predict much lower loads than beam theory,
- Stresses in the 84 inch pipe system may only be determined using finite element shell methods. Beam analysis can not be used.
- By determining the primary loading of the bends, appropriate conservative flexibility factors may be chosen to use in the pipe model.

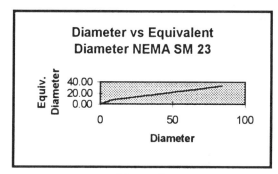

Figure 3

COMPRESSOR EXTERNAL LOADS

The compressor is designed to API 618 [4] which requires the flanges to withstand 1.85 the values calculated in accordance with NEMA SM 23 [5], which is the standard for Steam Turbines. This is usually given to the piping engineer as the limit to the external piping loads at the flange.

The allowable loads given for a 42 inch suction nozzle are relatively small compared to the flexibility's of the attached 84 inch pipe. The problem is to allow for small translations without bending the pipe, but remove the bending moments from the 100 foot overhead line from the scrubber.

The allowable loads are proportional to the diameter of the piping attached but are limited above 8 inch pipe. NEMA 23 equivalent diameters are shown in Fig. 3. The break in the curve shows the proportional decrease in allowable loads for large as compared to pipe 8 inch and under.

Expansion loops in an 84 inch pipe is pro-

Figure 4:photo of moment restraint

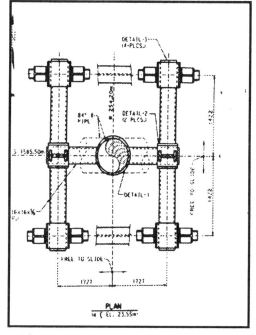

Figure 5:plan view of moment restraint

hibitive because of the size, also expansion joints are prohibited in Ethylene service. The only practical solution is a restraint device that allows translations but restrict rotations, a moment restraint device.

119

NEMA Factor	1.85	Shaft Angle w.r.t. ISO +X Coordinate	90		Loading Case	OPE

Operation Case

Nozzle	Diameter	Relative Position Vector			Load						Resultant	
		Rx	Ry	Rz	Fx	Fy	Fz	Mx	My	Mz	Fr	Mr
	mm	mm	mm	mm	N	N	N	N-m	N-m	N-m	N	N-m
Discharge	1067	1396	0	-1321	-202	1140	-50	143	0	911	1159	922
Node 110					-202	1140	-50	1649	337	2502		
Suction - 1	1067	0	0	0	1214	1461	988	-653	4034	980	2141	4202
Node 70					1214	1461	988	-653	4034	980		
Suction - 2	1067	2791	0	0	1296	-1019	2653	-5686	4383	998	3124	7248
Node 170					1296	-1019	2653	-5686	-3022	-1846		

Check combined RESULTANT forces and moments.

Calculated Diameters		Components of Combined Load						Combined Load		NEMA 23 & API 617		
		Fcx	Fcy	Fcz	Mcx	Mcy	Mcz	Fc	Mc	2Fc+Mc	250*Dc	
		N	N	N	N-m	N-m	N-m	N	N	English Units		
Equivalent	1848	2308	1582	3591	4690	1349	1636	4552	5147	5843	13992	42%
De (inch)	72.76	12447	31119	24895	18973	9487	9487					
Dc (inch)	30.25	19%	5%	14%	25%	14%	17%					

6

Caesar II Output

NODE	Case	FX	FY	FZ	MX	MY	MZ
110	OPE	-50	1140	202	911	0	-143
	SUS	-1371	-4501	365	85	0	-3930
70	OPE	988	1461	-1214	980	4034	653
	SUS	1745	-1940	-51	2229	-1194	2598
170	OPE	2653	-1019	-1296	998	4383	5686
	SUS	2145	-2087	-144	2280	-813	4191

Figure 6:NEMA 23 load reconciliation

MOMENT RESTRAINT DEVICE

Moment Restraint Devices are described in Kieth Escoe's book on process piping [1]. There are many variations on such a device. The particular design of the one used in this case was a function of the degrees of motion to be constrained as well as the particular physical restrictions of this application. Underground pipe running north and south prohibited any substantial structure in the north and south direction, which is the direction of the overhead pipe. The motions are constrained only to the east and west of the 84 inch pipe riser. Figure 4 shows a photograph of the MRS.

The device is supported by the pipe. The moment restraint device only constrains the moments and not the translations of the pipe. The unique crossbar design allows horizontal translational movement at each cross bar junction away from the pipe. The whole structure is allowed to rise up with the growth of the 84 inch pipe. Molybdenum Disulfide is used to lubricate the bearing surfaces which is a non sticky dry lubricant that offers a coefficient of friction less than .1. Figure 5 shows the plan view of the MRS.

LOCAL STRESSES AND RIGIDITY

In the analysis of the effectiveness of this device, the rigidity of the 84 inch pipe at the restraint location as well as the rigidity of the structure which transmitted the devices loads to the ground were considered. A stiffness was assumed at the external structures connection to the moment restraint device, and a shell finite element analysis was performed on the moment restraint and piping interface. In parallel, the structural engineer provided the necessary stiffness and proved it with a beam finite element analysis.

FINITE ELEMENT ANALYSIS

The finite element analysis were conducted with FEPipe, version 2.9 [7], constructed with 8 node, reduced integration , curved shell elements. The analysis considered thermal, internal pressure, full vacuum, and weight. The analysis was conducted using analysis methods shown in BPVC Section VIII, Division 2 [3] . Some of the details of the finite element shell model are shown in Appendix B, figures 1 through 6.

NEMA 23 FLANGE LOADS

In the development of this analysis, many NEMA SM 23 analysis were performed. Early on an automated system was developed to transfer results from the pipe stress analysis output to a NEMA spreadsheet shown in Fig. 6. The time consuming check of loads which involved transfer of coordinate systems and units was condensed to a five minute operation.

Visual Basic for Excell was used for automating the process. Loads were taken directly from the pipe stress beam analysis program and inserted in the spread sheet for units conversion and NEMA 23 analysis.

CONCLUSIONS

Compressor suction problems are difficult for small diameter pipe. For very large pipe the difficulty is magnified. Many things must be considered.

1. The real flexibility's of both pipe and structure must be considered for any realistic solution.

2. Large diameter pipe with D/t ratio's greater than 100 cannot be analyzed with beam finite element accurately, beam analysis may be used with shell element finite element analysis used to adjust the flexibility. Pipe stresses however are localized and must be obtained from the shell finite element analysis.

3. Friction plays an important roll in the analysis and must be considered accurately.

4. Many disciplines and skills must come together for the successful completion of such a task as this. This analysis came together with the skills of structural engineers, pipe support specialist, pipe and structural design experts, piping engineer, and finite element shell specialists. Data for the design came from many project design support team members.

5. New software tools such as programmable spreadsheets can ease the drudgery of an analysis and greatly improve the accuracy with the electronic transfer of data.

6. The appropriate use of an innovative design idea such as the moment restraint device.

ACKNOWLEDGMENTS
The author would like to thank Kerry Lee and Emer Posadas of Brown & Root, who detailed the structure and pipe support to exacting design constraints. In addition to Fred Hendrix of Paulin Research Group, who assisted in the finite element analysis.

References:

---

[1] Mechanical Design of Process Systems, Vol. 1 by A. Kieth Escoe 1986
[2] ASME B31.3 American Society of Mechanicl Engineers, Process Piping
[3] ASME American Society of Mechanicl Engineers, Boiler & Pressure Vessel Code
[4] .API 618 American Petroleum Institute, publication, Reciprocating Compressors for General Refinery Service.
[5] NEMA SM 23 National Electric Manufacturers Association, standard publication, Steam Turbines for Mechanical Drive Service.
[6] Caesar II proprietary pipe stress software by COADE.
[7] FEPipe pipe and vessel shell element finite element program by Paulin Research Group and ALGOR Inc.

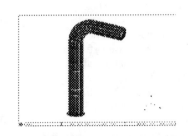

APPENDIX A

FINITE ELEMENT SHELL DERIVED FLEXIBILITY'S (f) AND STRESS INTENSIFICATION FACTORS (SIF)						
Direction	FEM Mvmt inch or °	FEM Stress psi	Beam Mvmt inch or °	Beam Stress psi	flexibility	SIF
+X	4.031	21064	.0112	103	360	205
-X	-3.929	20937			351	203
+Y	.0238	3150	.0018	30	13.2	105
-Y	-.0181	3133			10.0	104
+Z	.0274	4013	.0206	95	1.33	42.2
-Z						
+ØX	.0519	38571	.0094	373	5.52	103
-ØX						
+ØY	.0165	4041	.009	297	1.83	13.6
-ØY						
+ØZ	.00693	3939	.002	122	3.46	32.2
-ØZ	.001976	3795			1.	31.1

Table 1

Direction	Reference to base		Direction	Reference to base
x	Shear in plane		rx	Bend out plane
y	Tension		ry	torsion
z	Shear out plane		rz	Bend in plane

APPENDIX B

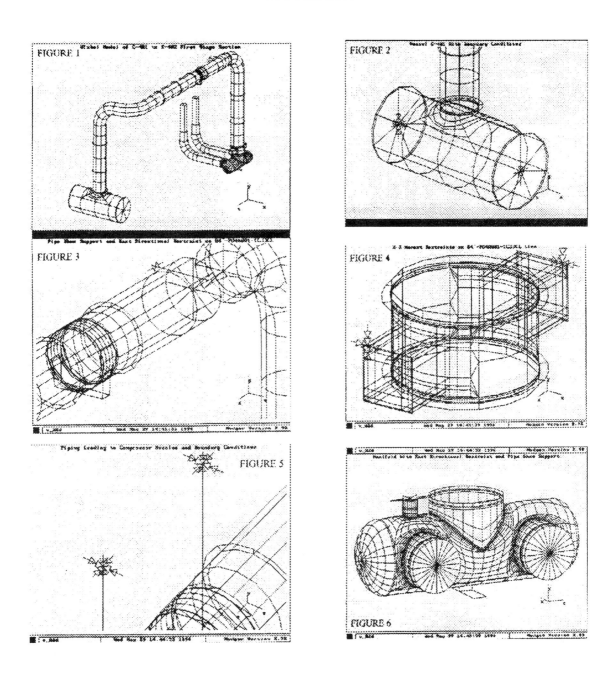

COMPONENT ANALYSIS AND EVALUATION

Introduction

L. Ike Ezekoye
Westinghouse Electric Corporation
Pittsburgh, Pennsylvania

Papers and presentations in this session, Component Analysis and Evaluation, cover areas on valve design, testing, analysis and evaluation. While the valve industry is a mature industry, there remain a number of areas where much needs to be learned to improve valve performance, testing and maintenance. In this session, there are two papers and two oral presentations.

The paper by *Ezekoye, Moore, Gore, Dandreo, and Turner* discusses the instability that may arise in piloted valves subjected to two phase flows. The authors show that orifice size differences can make the valve's performance unpredictable. . The paper by *Ezekoye, Legenzoff, and Walker* presents the strategies for maintaining body-to-bonnet flanges of valves. It shows that these rather mundane issues have no easy solutions.

Bunte's oral presentation plans to address the potential for diagnostic testing of motor operated valves from the motor control center. The presentation is topical as there is ongoing regulatory interest in condition monitoring of safety related motor operated valves in the nuclear industry. Finally, *Hopkins'* oral presentation discusses the merits of MagIon coating for valve parts such as bolting and studs to reduce friction.

PVP-Vol. 356, Integrity of Structures and Fluid Systems,
Piping and Pipe Supports, and Pumps and Valves
ASME 1997

CHARACTERIZATION OF ORIFICE PERFORMANCE IN MIXED FLOWS

L. I. Ezekoye, P.E.
W. E. Moore
Westinghouse Electric Corporation
Pittsburgh, Pennsylvania

J. J. Gore, P.E
D. Dandreo
Wolf Creek Nuclear Operating Corporation
Burlington, Kansas

A. Turner
Dominion Engineering
McLean, Virginia

ABSTRACT

In some Westinghouse pressurized water reactors (PWRs), pilot operated solenoid valves are used to provide overpressure protection. The valves are hydraulically assisted devices which depend on the existence of differential pressure across the plug to open and close. To both create the differential pressure and regulate the opening and closing speeds of the valves, orifices are used in series. Optimal sizing of the orifices assures the operational reliability of the valves. Generally, orifices are sized for a single phase flow (liquid or vapor). Mixed flows may result in erratic behavior depending on the ratio of the orifice areas. This paper presents a case study where mixed flows affected the performance of the two orifices in series which are employed in the design of pilot operated solenoid valves. The paper also presents a proposed solution to improve stability and reliability.

NOMENCLATURE

A_1	=	Inlet orifice area
A_2	=	Outlet orifice area
C_p	=	Constant pressure specific heat
C_v	=	Constant volume specific heat
C_2	=	Sonic velocity at the outlet orifice (1653 ft/sec for steam at 900°R)
G	=	Mass flux
g	=	Acceleration of gravity (32.174 ft/sec^2)
k	=	Specific heat ratio, C_p/C_v
K_m	=	incipient cavitation index
M_1	=	Mass flowrate through the inlet orifice
M_2	=	Mass flowrate through the outlet orifice
P_t	=	Minimum (threshold) differential pressure required to open the valve
P_v	=	Vapor pressure at the temperature (psia)
P_1	=	Inlet absolute pressure (psia)
P_2	=	Chamber absolute pressure (psia)
R	=	Ideal gas constant (85.76 ft-lbf/lbm-°R for steam)
r_c	=	Critical pressure ratio
T_1	=	Inlet fluid absolute temperature
T_2	=	Chamber fluid absolute temperature
V_1	=	Flow velocity across the inlet orifice
V_2	=	Flow velocity across the outlet orifice
X	=	Percent water in a saturated mixture
ΔP_1	=	Differential pressure across the inlet orifice
ΔP_2	=	Differential pressure across the outlet orifice
ρ_1	=	Water density at the entrance to the inlet orifice
ρ_2	=	Water density in the chamber between the two orifices
ρ	=	Fluid density

INTRODUCTION

In Westinghouse PWRs, the power operated relief valves (PORVs) serve a number of functions, depending on the plant licensing basis. Some of the key functions are:

- provide cold shutdown capability following a severe accident such as loss of feedwater
- mitigate overpressure transients that might otherwise result in a plant trip
- minimize challenges to the safety relief valves
- depressurize the primary circuit in the event of a steam generator tube rupture
- fill and vent the pressurizer
- provide low temperature over-pressure protection of the reactor vessel

The solenoid power operated pressure relief valve is an electrically controlled, pressure assisted globe valve. Figure 1 is a sketch of the PORV used on the reactor coolant system pressurizer. Basically, the PORV incorporates a pilot valve in conjunction with the system pressure to open and close the main valve.

In order to understand how the seating loads are developed, it is necessary to describe in some detail the PORV design. Figures 2a and 2b are simplified schematics of the valve showing how the pilot valve communicates with the main valve to cause it to open and close. Referring to Figure 2a, with the pilot valve closed (de-energized), the control chamber pressure builds up to the inlet pressure. The resulting pressure differential across the main seat keeps the main valve plug closed. With the pilot valve open (energized), the pressure in the control chamber drops to the level of the downstream pressure. An upward pressure differential is imposed on the plug and the valve opens. Unlike an air operated globe valve, where the plug is coupled to an operator to open and close the main port, the motion of a piloted valve plug depends on the direction of the pressure differential across the plug. The feed orifice controls the closing and the bleed orifice controls the opening. The solenoid valve port is considerably larger than the bleed orifice and does not affect the valve opening. In the remainder of this paper, the closing (feed) orifice will be considered closed and having no effect on the opening of the valve. The opening (bleed) orifice will be considered in series with an equivalent orifice representing the gaps at the plug and cage piston rings. The equivalent orifice and the opening orifice will together control the opening of the main valve.

WOLF CREEK PORV

The Wolf Creek Pilot Operated PORVs are as described above. The design specification for the PORVs indicates a minimum required design operating pressure of 420 psig during Cold Overpressure Mitigation System

(COMS) operation. These valves are designed to limit the pressurizer pressure to a value below the high pressure trip set point for all design load reduction transients up to and including a 50 percent load reduction with steam dump actuation. The relief valves will operate automatically or they may be operated remotely from the control room. The valves are designed to prevent challenging of the spring loaded safety valves under certain transients.

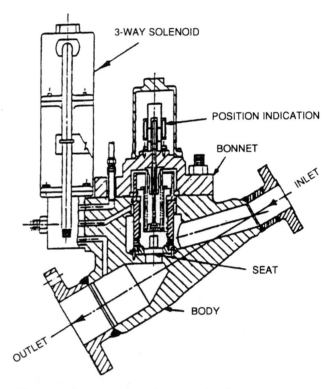

Figure 1. Cross-Section of Reactor Coolant System PORV

HISTORY

During plant cool-down on February 5, 1996 for Refueling 8 (RF8), the Wolf Creek operators attempted to cycle the PORVs in accordance with plant procedures. One of the valves was actuated seven times and the other valve was actuated three times. The PORVs did not respond in the way that was expected. The position indication device for one valve flickered during the first attempted actuation. During subsequent actuations of the same valve, no change in the position indication was observed. The position indication of the other valve showed that the valve opened and closed as expected during the first actuation but did not register valve opening during the subsequent actuations.

EVALUATION

A three pronged approach was undertaken to develop a root cause for the faulty operation of the Pressurizer Power Operated Relief Valves. This included:

1. mechanical inspection of the valves to determine if any internal binding was occurring,
2. inspection of the solenoid valve to confirm that the solenoid valve is not leaking excessively and that the solenoid is functional, and
3. a systems evaluation to confirm whether or not the valves actually operated.

Mechanical Inspection

The results of the mechanical inspection of the valve internals confirmed that the valve parts were within the manufacturer's tolerances. No evidence of binding was observed. Thermal binding was ruled out as a probable cause.

Solenoid Valve Inspection

The solenoid valves were inspected and tested for excess seat leakage and were found to be within the design requirements. Inadequate performance of the solenoid valves was ruled out as the probable cause.

System Evaluation

Since the PORVs depend on differential pressure to operate, the first approach was to confirm that adequate pressure was available in the system to operate the valves. A review of the system conditions during the period when the valves were actuated confirmed that the minimum required actuation pressure was present during the event. There was ample differential pressure across the valves to operate them successfully.

Next, the operability of the position indication devices was explored. This was done by reviewing the recorded system perturbations along with the data collected from previous valve relievings to identify if the valves operated. For this review, changes in system pressures, temperatures and volumes were compared against each other to identify whether valve motions occurred during the event to perturb the system. The argument considered here is that, if the system and the valves are well behaved, changes in the system due to valve operation should be consistent. Consistency provides confidence in saying that the valves opened, and the converse is obvious.

The results of the system evaluation showed that the position indication devices were operable. However, the valves failed to open fully several of the times they were actuated and failed to open at all in some cases. It was conjectured that the increased stroke time for these actuations may have been due to the fact the valves were being operated at off-design conditions. The investigation, therefore, focused on determining the performance of the valves in an off-design condition that could have affected their response.

PERFORMANCE OF THE PORV UNDER DIFFERENT FLOW MEDIA

The PORVs have been tested successfully in both steam and water using the manufacturer's specified orifices. During the RF8 stroke testing, plant operations reported that there was simultaneous operation of the pressurizer spray system, pressurizer heaters and venting of the pressurizer via the PORVs occurring which would have resulted in a two phase flow through the PORV. The performance of the orifices during that mode of operation was explored next.

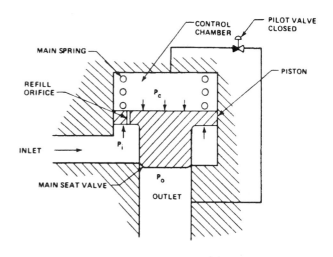

Figure 2a. External Piloted Valve (Closed Position)

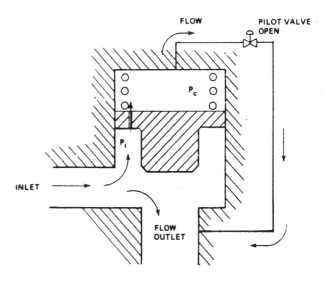

Figure 2b. External Piloted Valve (Open Position)

AN ASSESSMENT OF THE PERFORMANCE OF TWO ORIFICES IN SERIES

Flow in a piping system with two orifices in series may be considered as an idealized model of the plug and bonnet chamber during opening of a pilot operated PORV. See Figure 3. The ratio of the inlet orifice area (A_1), representing the gaps at the plug and cage piston rings, to the outlet (opening) orifice area (A_2) is 0.5102. The performance of two orifices in series with four different flow media is examined next. The first case that will be considered is steam flow. The second case is the flow of subcooled water. The third case is the flow of saturated water, and the last case is two phase flow.

For these four cases, we will examine

(a) whether there exists a mass flow imbalance between flow through the inlet orifice and flow through the outlet orifice such that the chamber begins to repressurize, and

(b) if the differential pressure across the plug is greater or less than the minimum threshold pressure required to open the valve.

Symbolically, the governing criteria for valve performance for all four cases are that the valve will open if and only if

$$M_2 \geq M_1 \text{ and}$$

$$P_1 - P_2 \geq P_t$$

ANALYSIS CONDITIONS

The analysis fluid conditions are:

Saturated Steam/Water Conditions Inlet Pressure = 380 psia
Subcooled Water Conditions   Inlet Pressure = 380 psia
                                 Inlet Temperature = 100°F

CASE 1 - STEAM FLOW

Assume that both orifices are choked. This assumption provides a bounding case. Thus the ratio of the mass flow rates through the two orifices is [Reference 1]:

$$\frac{M_1}{M_2} = \frac{A_1}{A_2} \frac{P_1}{P_2} \sqrt{\frac{T_2}{T_1}}$$

The above equation assumes that dry steam obeys ideal gas laws. Thus, the ratio of the total pressures is:

$$\frac{P_2}{P_1} = (\frac{2}{k+1})^{\frac{k}{k-1}}$$

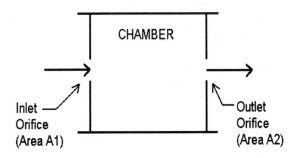

Figure 3. Idealized Model of Plug and Bonnet Chamber Inlet Orifice Represents Piston Ring Gap

where k = 1.1

Thus,

$$\frac{P_2}{P_1} = 0.5847$$

and

$$P_2 = 222 \, psia$$

From the steam table,

$$T_1 = 440° F = 900° R \text{ (saturation temperature at } 380 \text{ psia)}$$

$$T_2 = 390° F = 850° R \text{ (saturation temperature at } 222 \text{ psia)}$$

The ratio of the mass flow rates is then:

$$\frac{M_1}{M_2} = 0.5102 \bullet \frac{1}{0.5847} \bullet \sqrt{\frac{850}{900}}$$

$$\frac{M_1}{M_2} = 0.848$$

130

Thus, with both the inlet and the outlet orifices choking, the out-flow is greater than the in-flow. The differential pressure across the plug is 158 psi = (380 psia - 222 psia), which is more than P_t, the minimum differential pressure required to stroke the valve. The valve will open and relieve.

CASE 2 : COLD WATER CASE

For sub-cooled water case, we assume isothermal behavior.

The ratio of the mass flow rates through the first and the second orifice is:

$$\frac{M_1}{M_2} = \frac{\rho_1 A_1 V_1}{\rho_2 A_2 V_2}$$

By definition, $\rho_1 = \rho_2 = \rho$

$$V_1 = \sqrt{\frac{2 g_c \Delta P_1}{\rho}}$$

$$V_2 = \sqrt{\frac{2 g_c \Delta P_2}{\rho}}$$

and

$$\frac{V_1}{V_2} = \sqrt{\frac{\Delta P_1}{\Delta P_2}}$$

We assume the orifices are cavitating. For cavitating orifices,

$$\Delta P_1 = K_m (P_1 - r_c P_v)$$

and

$$\Delta P_2 = K_m (P_2 - r_c P_v)$$

Since the fluid temperature is low (100°F) the vapor pressure is ignored. Hence,

$$\Delta P_1 = K_m P_1$$

and

$$\Delta P_2 = K_m P_2$$

But the pressure in the chamber is related to the inlet pressure as follows:

$$P_2 = P_1 - \Delta P_1$$

Thus,

$$P_2 = P_1 (1 - K_m)$$

and

$$\Delta P_2 = K_m (1 - K_m) P_1$$

Substituting the above expression into the mass flow rate ratio, we have:

$$\frac{M_1}{M_2} = \frac{A_1}{A_2} \cdot \sqrt{\frac{1}{1 - K_m}}$$

The value of the incipient cavitation index is generally experimentally determined. From Reference 2, the value of K_m ranges from 0.264 to 0.35.

Substituting, we have, for $K_m = 0.264$:

$$\frac{M_1}{M_2} = 0.595$$

$$\Delta P_1 = K_m P_1 = 100 \, psi$$

For $K_m = 0.35$:

$$\frac{M_1}{M_2} = 0.633$$

$$\Delta P_1 = K_m P_1 = 133 \, psi$$

The pressure available to open the valve is ΔP_1.

Hence, $133 \, psi \geq \Delta P_1 \geq 100 \, psi \geq P_t$.

Thus it is seen that, for relieving subcooled water, the PORVs should work well.

CASE 3 : SATURATED WATER RELIEVING

In the saturated water case, it is assumed that the plug/bonnet chamber is initially filled with steam. However, saturated water is below the plug such that as soon as the valve is signaled to open, the saturated water begins to enter the chamber via the piston ring orifice. At the same time, the opening orifice is venting the steam from the chamber. This is the condition that is being analyzed.

The inrush flow is:

$$M_1 = G \bullet A_1 \bullet \rho_1$$

The mass flux, G, is from Reference 3:

$$G = 21.19 \bullet P_1^{0.861} \; \frac{lbm}{ft^2 \bullet sec}$$

The out-rush flow is steam and is defined as follows:

$$M_2 = A_2 \bullet C_2 \bullet \rho_2$$

where C_2 is the sonic velocity defined as follows:

$$C_2 = \sqrt{kgRT_2}$$

Then

$$\frac{M_1}{M_2} = \frac{G \bullet A_1 \bullet \rho_1}{A_2 \bullet C_2 \bullet \rho_2} = \frac{A_1}{A_2} \frac{G \bullet \rho_1}{C_2 \bullet \rho_2}$$

Substituting values into the equation, we have

$$\frac{M_1}{M_2} = 1.324 > 1$$

This indicates that the opening orifice chokes. When the opening orifice begins to choke, the chamber pressure builds up and keeps the valve closed.

CASE 4 : MIXTURE OF SATURATED WATER AND STEAM

In Case 3, it is noted that relieving saturated water with the designed orifice may lead to valve instability. In this case, we are interested in finding out what happens when a mixture of water and steam is relieved.

Here, we assume that saturated water occupies X% of the inlet orifice area and the steam occupies the rest.

From the saturated water case, the mass flow rate of the inrush saturated water fraction is:

$$M_{1W} = G \bullet A_1 \bullet \rho_1$$
$$M_{1W} = M_1$$

The inrush saturated steam flow, using the results of the previous (Case 3) calculation with adjustments for the areas A_1 and A_2, is:

$$M_{1S} = (100 - X) \bullet \frac{A_1}{A_2} \bullet \frac{M_2}{100}$$

For the valve to have a chance to operate,

$$M_{1W} + M_{1S} = M_2 \; (Case \; 3)$$

or, substituting,

$$\frac{X}{100} = \frac{1 - \dfrac{A_1}{A_2}}{\dfrac{M_1}{M_2} - \dfrac{A_1}{A_2}}$$

and

$$X = 60$$

The analysis shows that a mixture of steam and water consisting of 40% steam and 60% water by volume will result in the valve not working properly. Considering the model simplification, it is prudent to say that a water/steam mixture of 50%:50% can create unstable flow dynamics inside the valve.

TECHNICAL INSIGHT

The opening time of the PORV is, by design, governed by the size of the opening orifice. In principle, during valve opening, the plug/bonnet chamber is connected to the downstream piping via the opening orifice. Evacuation of the pressure in the chamber permits the inlet pressure acting underneath the plug to lift the plug and open the valve. By design, the chamber receives leakages across the piston rings (cage and plug) during all relieving. Thus, one can idealize the flow through the chamber as flow across two orifices in series. The first orifice is formed by an equivalent orifice associated with the plug/piston ring gaps. The second orifice is the opening design orifice. Tests performed during hot functional testing (HFT) with steam and subcooled water indicate that the first orifice is the limiting (choked) orifice under these conditions. By this being the limiting orifice, the chamber pressure is less than the inlet pressure and sufficient differential pressure is developed across the plug to open the valve. The analysis confirms this. If, on the other hand, the valve is exposed to a mixture of water and steam (2 phase flow), the flow past the first orifice (i.e., the leakage) flashes in the chamber and repressurizes the chamber, thus making the second or opening orifice the flow limiting device. The mass input to the chamber from piston leakage is higher than the mass output through the opening orifice.

The mass left in the chamber repressurizes the chamber. Under such a condition, only a small differential pressure exists across the plug and this slows down or degrades the valve movement. When such a choking condition occurs, the valve may go into a hammering operation (i.e., chattering) at the seat because of repeated small openings and closings.

From the foregoing analysis, a practical way to minimize mass imbalance across the chamber at the off-specification operating conditions is to enlarge the opening orifice. Enlargement of the opening orifice is consistent with Reference 4, which shows that for such systems (i.e., two orifices in series), the limiting orifice is always the inlet orifice if the area ratio of the outlet orifice area to the inlet orifice area is ten, regardless of the pressure ratio.

CONCLUSION

This paper has examined the performance of the PORV under four relieving conditions. It shows that there are no performance problems under pure steam conditions and under subcooled water conditions. However, as the water relief approaches saturated conditions, the ability of the opening orifice to relieve is doubtful if its area is not sufficiently larger than the area of the orifice representing the plug/piston ring gap. A mixture of steam and water to the level of 50% water and 50% steam creates an unstable flow imbalance inside the valve chamber which degrades the valve stroke speed. One way to alleviate the flow imbalance is to open up the size of the opening orifice and thus improve the valve reliability and performance. It should be pointed out, however, that if the opening orifice is removed or enlarged, the relief line piping will require requalification to account for higher discharge loads due to faster valve opening times.

REFERENCES

1. Howell, Glen W., and Weathers, Terry M., editors. *Aerospace Fluid Component Designers Handbook, Vol. I, Revision B.* Air Force Rocket Propulsion Laboratory, RPL-TDR-64-25, March, 1967.

2. Stuart, Milton C., and Yarnall, D. Robert, "Flow Through 2 Orifices in Series," Special Research Committee on Fluid Meters, ASME Semi-Annual Meeting, June 25-30, 1936.

3. Benedict, Robert P., "Maximum Flashing Flow," *Fundamentals of Pipe Flow.* John Wiley & Sons, Inc., New York, 1977, pp. 346-347.

4. Andersen, Blaine W., *The Analysis and Design of Pneumatic Systems.* John Wiley & Sons, Inc., New York, 1967.

PVP-Vol. 356, Integrity of Structures and Fluid Systems,
Piping and Pipe Supports, and Pumps and Valves
ASME 1997

ON-LINE MAINTENANCE OF VALVE BOLTED BODY-TO-BONNET JOINTS

L. Ike Ezekoye, P.E
T. J. Legenzoff
Westinghouse Electric Corporation
Pittsburgh, Pennsylvania 15230

Herbert P. Walker, P.E
Southern Nuclear Operating Company, Inc.
Birmingham, Alabama

ABSTRACT

The majority of maintenance activities on valves in Nuclear Power Plants are performed during outages. A large portion of these maintenance activities involves either seat leakage or body-to-bonnet leakage. While it is recognized as difficult for plants to perform seat repairs on-line, there is an increasing interest in on-line maintenance of valve body-to-bonnet connections. This interest is driven by the desire of plants for longer operating cycle times and for shorter and shorter refueling outages. This paper discusses the various technologies available for on-line maintenance and the technical issues that need to be resolved to make each technology viable.

INTRODUCTION

Valves constitute the most common equipment in nuclear power plants. Their large population is one of the reasons why valve failures are major contributors to the list of power plant forced or scheduled outages. See Table 1. Typical valve failures (Table 2) include external leakage, internal leakage, failure to open and failure to close. Of all valve operational problems, leakage (external and internal) is the most prevalent. Both external and internal leakage waste valuable resources (pressurized fluid), damage the equipment and pose a personnel hazard. As such, they are both undesirable. While it is recognized that packing leakage remains a significant contributor to external leakage, except for unusual situations of packing blowout, plants have at their disposal a number of emergency measures such as back-seating and retorquing to address urgent valve packing leakages. This paper focuses on external leakage from valve body-to-bonnet joints for the reason that it is one failure which a plant operator intuitively feels can be readily solved without bringing the plant down. Even if emergency shutdown is not the driver for on-line repair, there are other concerns such as safety of both personnel and adjacent equipment that dictate that immediate actions be taken. In nuclear power plants, for example, leaking flange connections are a major source of radioactive contamination of plant personnel and equipment. Increased radiation levels can increase the mean-time-to-repair of the leak or surrounding equipment.

Table 1

1989-1993 PWR Top System/Component Cause Codes Forced and Scheduled Outages and Deratings

Component Cause	No. of Occurrences per Unit-Year
Condenser Leaks	2.06
Refueling	0.99
RCP's & Motors	0.77
SG Leaks/Tubes	0.49
Valves	0.32
Unit Performance Testing	0.25
BOP Misc.	0.24
Core Washdown	0.23
High Pressure Injection	0.17
RCS Problems (other)	0.13
Control Rod	0.10
Source: Generating Availability Report 1989-1993; "Availability Performance of Electric Generating Units in North America", June 1994, North American Electric Reliability Council	

Table 2

Valve Failure Rates Valve Failures Reported in NPRDS from 7/74 to 11/92 Gate Valves, Check Valves and Globe Valves

Failure mode	Failures/Hr
External Leak	6.60E-06
Internal Leak	5.37E-06
Failure to Operate as Required	2.80E-06
Failure to Close	1.92E-06
Failure Found During Testing	1.70E-06
Premature Opening	1.69E-06
Failure to Open	8.96E-07
Failure to Remain Open	4.87E-08

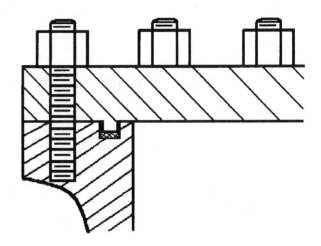

Figure 1A
Typical Body to Bonnet Flange Joint Without Seal Weld Provision

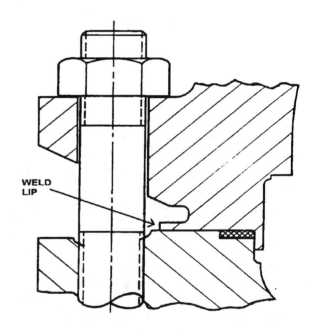

Figure 1B
Body to Bonnet Flange Joint with Seal Weld Provision

BODY-TO-BONNET DESIGN AND FAILURE MECHANISMS

The design of valve body-to-bonnet joints and the mechanisms by which they fail are not the main topics of this paper. However, it is felt that reviewing their design and how they fail would be useful in understanding the effectiveness of different emergency solutions to valve flange leakages. Figures 1a and 1b show two cross-sections of body-to-bonnet joints. Figure 1a shows a design with seal weld provisions while Figure 1b shows a design which is not intended for seal welding. There are a wide variety of commercially available gasket materials for different applications. Typical gasket materials include asbestos, graphite composites, soft metals, Teflon, etc.

The design of valve body-to-bonnet flanges is vendor specific. No two designs are alike. Each valve vendor maintains its own standards taking into consideration the size

136

and pressure class of the valve, the fluid properties, the gasket manufacturer's recommendations, system operating conditions (pressures and temperatures), the expected life of the gasket and cost.

The ASME Boiler and Pressure Vessel Codes (Sections III and VIII) establish the criteria for flange analysis [1, 2]. These rules provide the means to assure structural integrity of the joint irrespective of the differences in the joint designs. The rules also consider the leak tightness of the joints. While the rules have served the industry well, failures continue to occur. The US Pressure Vessel Research Council (PVRC) has sponsored several research activities to provide appropriate gasket factors (m and y) to improve the leak tightness of joints using the ASME guidelines. Over the years a number of investigators have performed studies to improve the reliability of gasketed joints. Hsu and Payne provide an anthology of research programs sponsored by the PVRC [3]. The studies cover elevated temperature design, gasket materials, creep and relaxation behavior of gasket materials, gasket factors and so on. All these studies aim to improve gasket performance. However, leaks still occur and when they do, a number of emergency measures are taken.

There is no single reason why valve flanges leak. Over the years, various failure mechanisms have been identified and studied. The observed failure mechanisms include improper design and qualification of the joint, aging of the gasket material, material creep, thermal transients, misapplication, and improper assembly/maintenance. Table 3 summarizes the frequency of body-to-bonnet failure initiating mechanisms for different valves. These data were derived from the Nuclear Power Reliability Data System (NPRDS). As the table shows, the most common cause of valve leakage is wear and aging of the flange gasket. There is also a high incidence of unknown causes. Erosion/corrosion is the next most common identifiable cause of leakage followed by improper maintenance, and thermal and mechanical cyclical fatigue.

Table 3
Causes of External Leaks
Gate Valves, Check Valves and Globe Valves

Cause of Failure	%
Wear/Aging	54.6%
Unknown	13.7%
Insufficient Compression	7.9%
Improper Maintenance	7.3%
Other	7.3%
Erosion	3.1%
Cyclic Fatigue	3.1%
Design/Manufacturing Errors	2.1%
Body Deformation	0.8%

EMERGENCY REPAIR ACTIONS

Several emergency maintenance options exist for solving the problem of a leaking flange. The key options discussed in this paper are:

    (a)   retorquing
    (b)   seal welding
    (c)   seal capping
    (d)   seal injection
    (e)   freeze plugging with gasket replacement or with seal welding or seal capping

RETORQUING

Retorquing of bonnet studs is often the first emergency repair action to seal leaky joints. This action presumes that either the joints are improperly torqued or the gasket has relaxed. Retorquing is attractive because it is non-intrusive, fast and does not require bringing the system down. But retorquing has its limitations. It is not very effective except in a few isolated cases where the leakage is quickly detected. If leakage is allowed to proceed for any length of time, the effluent could wash away the gasket sealing material or erode the seal surface in the case of metal to metal seal rings. In these cases retorquing is not effective. In some cases, torque cannot be increased because of design limitations such as the allowable stress limits of either the stud, the body tapped hole or the flange. In cases where the joint is improperly designed to begin with, torquing is not effective. These areas need to be considered before retorquing takes place.

SEAL WELDING

As the name implies, seal welding involves the application of weld metal to the joint between the flange faces. See Figure 2. The weld metal forms a barrier to the leak path which has developed. Seal welding can be performed while the system is on-line provided that the system pressure can be reduced to a level which stops the joint from leaking. Otherwise, the system may need to be taken out of service, depressurized and drained or freeze plugged as with other permanent repairs. Freeze plugging, which is a technique to provide temporary isolation of the valve, will be discussed separately.

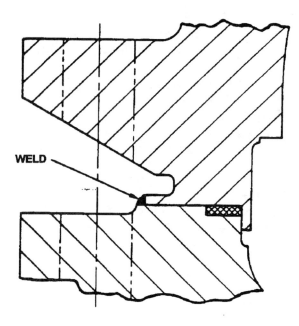

Figure 2
Flange Joint Seal Weld

In order to perform a seal weld repair, flange studs must be removed so the joint formed by the flange faces can be accessed. See Figure 1a. Generally, on-line seal welding requires removing one stud at a time to expose the mating surfaces. This, of course, is time consuming. However, the more studs that can be safely removed without jeopardizing pressure boundary integrity, the faster and more efficiently the repair can be completed. Unfortunately, removal of multiple studs imposes a considerable amount of risk. The process of removing studs, welding and replacement of studs during seal welding progresses circumferentially around the joint until the entire joint has been welded. The sequence and location of stud removal may be optimized to minimize the effort associated with this approach. This, of course, depends on the valve design and plant conditions.

Seal welding is an effective way to contain leakage. It is unique in that it is a permanent solution that can be completed while the valve is in service. As such, the repair is cost-effective relative to other repair approaches which are either not permanent or cannot be completed on-line. Evaluations performed to support the repair usually ensure that the repair is performed safely and efficiently.

The main drawbacks of seal welding include the fact that many valves are not designed with a seal welding feature and that, in large leaks, it is exceedingly difficult to make the seal weld. Additionally, on-line seal welding often requires isolating and depressurizing the line. Where total depressurization is not possible, an evaluation is often necessary to confirm joint integrity during the welding process. Another major drawback of seal welding is the level of resource intensity associated with it. Seal welding is difficult to perform and personnel risk is high. For example, in order to assess the pressure integrity of the joint as studs are removed, detailed knowledge of the joint design is needed such that necessary stress and leak worthiness calculations can be performed. Flange and stud stresses with (n-x) studs are calculated and checked to confirm that the flanged joint will maintain pressure boundary integrity for all design basis operating modes. If it has been determined that the joint has been leaking for some time, the designer must determine if the stud diameter should be reduced for the calculations to account for potential corrosion of the stud material. In addition to checking flange and stud stresses, the designer should confirm that adequate gasket compression is maintained at all times to prevent gross leakage and potential injury to workers. For these reasons, it is generally advisable to involve the valve manufacturer during a seal welding process to provide support.

Seal welding can only be performed if the existing flange has provisions for seal welding. As discussed earlier, there is greater personnel risk than with some other approaches such as retorquing. From field experience, stud removal can be challenging and the potential exists for damaging the studs and/or tapped holes. Additionally, if the joint must be opened at some time in the future for internal repair or inspection, a greater effort will be required to separate the flange faces than for any other approach.

SEAL CAPS

Seal caps are, as the name implies, caps permanently attached to valves to contain the effluent. Basically, they encapsulate the leaky joint. By design, a seal cap is a non-pressure boundary attachment to the valve. Figure 3 is an illustration of one of these devices installed on a check valve. It is a fabricated shell welded to the valve body to provide a backup seal. The seal cap is very effective in containing external leakage. However, it has its limitations. First, it requires welding to the valve and possibly to the bonnet/cover. Next, the studs and nuts are exposed to the leaked fluid which may, depending on the stud material, stud stress level and the fluid conditions (chemistry and temperature), promote intergranular stress corrosion cracking or corrode the bolting materials. A seal cap is bulky and adds weight to the valve. The additional weight may require a reevaluation of the valve for structural integrity and natural frequency and a review of the piping for adequacy of existing qualification. The bulkiness of seal caps tends to restrict their application to small valve sizes. Also, with very large leaks, making a body weld may be impossible, thus requiring depressurization of the system. Finally, removal of the seal cap for permanent maintenance increases the risk of damaging the body or the bonnet.

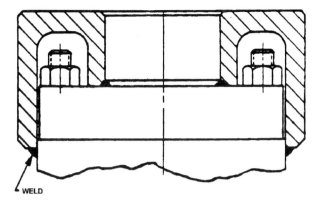

Figure 3
Seal Cap Cross-Section

SEAL INJECTION

Seal injection is the most frequently used emergency approach to manage a leaky valve flange and it is the least accepted in the primary side of Westinghouse Pressurized Water Reactors (PWRs). Table 4 lists the frequency of use of seal injection processes and other approaches. The basic principle of seal injection is to form a close toleranced specially designed barrier around the leak into which a sealant is injected at high pressure to form a back-up seal. Figure 4 is an illustration of such a process. Other variations exist in the industry and are generally custom built to suit the specific problems at hand. Essentially, seal injection is a special case of the seal cap without the welding and with a sealant injected into the enclosure. The process has the advantage of sealing a leak without shutting down the system.

Table 4

Distribution of Plant Actions Following Body-to-Bonnet Leaks for Valves Greater than 3 Inches, 1989 - 1994

Actions	%
Seal Injection	49.2%
Seal Repair*	48.4%
Left to Leak	1.5%
Shutdown	0.8%
* - Includes Seal Weld, Freeze Plug and Replacement	

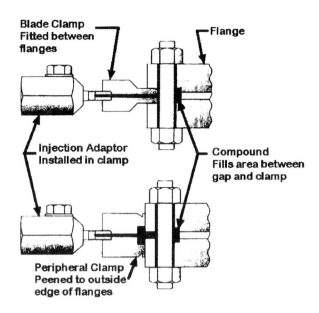

Figure 4
Sealant Injection

Source: Furmanite Bulletin "On-Line Leak Sealing"

The sealants are thermosetting organic compounds specifically formulated to cure at defined temperatures. The choice of sealants for a specific application depends on the process fluid properties, system conditions, materials of construction of the valve and other components that might be in contact with the sealant. A number of sealants are available commercially for different applications. The sealant suppliers are usually responsible for designing and installing the sealant barriers and for injecting the sealants. It is generally advisable to involve the valve manufacturer during seal injection if any drilling of the valve is required.

The application of sealants requires fabrication of a dam around the joint followed by seal injection. Typical barriers are bulky which add weight to the valve and piping. In general, an analysis is required to confirm that the structural integrity of the joint is not impaired and that the natural frequency of the valve is not affected. Additional analysis is usually performed for the qualification of the piping.

The main advantage of this process is that it can be performed without shutting down the plant. Major concerns about this process exist, however. The chief one is that vendors have yet to qualify sealants under strict quality assurance standards for the PWR primary system environment. Also, the method of application (i.e., injection process) increases the likelihood of sealants entering the system and contaminating it. Thirdly, the sealants are for one thermal cycle. Thus, thermal cycling reintroduces leaks.

Another concern is that sealants have a number of contaminants in their formulation such as chlorides, fluorides, sulfur, lead, etc. which are either undesirable in a radioactive environment or can attack the valve materials. Also, the removal of sealants on valve surfaces during disassembly for permanent repair is generally time consuming. Finally, the weight issue, while it is addressable, needs to be evaluated to confirm that existing valve and piping qualifications are not adversely affected.

FREEZE PLUG AND GASKET REPLACEMENT

In some cases where a system can be isolated for a relatively short time for work to be performed on a valve and no means have been provided for isolating it, freeze plugging is used to isolate the system. Freeze plugging is a process which involves providing an isolating plug to a leaky valve to permit maintenance to be performed on the valve. It is a non-intrusive technique in which liquid nitrogen is injected into an externally mounted jacket to freeze the contents in the pipe to form a plug. The time to freeze the contents depends on the pipe size, fluid properties and system temperature. See Table 5 for one vendor's recommended freeze times. Following freezing and subsequent leak repair, the nitrogen jacket is removed and the thawing proceeds naturally.

The freeze plugging process is an effective means of gaining access to the valve for a permanent repair or replacement of a gasket. It is a means to an end, the end being the stoppage of a leak. Freeze plugging can be used in performing any of the other measures such as capping and seal welding. It is generally more expensive than any of the other approaches. However, it can be used for a short repair window.

REMARKS

The above discussions address the most common emergency solutions employed in power plants to address valve body-to-bonnet leaks which are undesirable for economic and environmental reasons. While technologies exist for emergency situations, the available technologies have differing levels of effectiveness and different issues associated with them. Retorquing the body-to-bonnet studs is inexpensive but is marginally useful. Seal welding is effective but is limited to valves with seal weld capability. Seal caps are equally effective but are best suited for check valves and for valves constructed with parts not affected by the effluent. Seal injection is very effective but sealants compatible with the primary system chemistry and materials of construction are yet to be qualified. Finally, freeze plugging is an effective means to isolate a line to permit a permanent repair of the leak while the system is on line but it is expensive. From a plant operator's perspective, it appears that the best promise for future emergency sealing lies in developing and qualifying new sealants for use in the primary side of PWRs. Developing new sealants for primary and secondary chemistries would provide backup choices in an area such as valve body-to-bonnet leakage where no solution is optimal.

ACKNOWLEDGMENT

The authors extend their thanks to Southern Nuclear Operating Company for stimulating this investigation and to Terry Matty, Edward Petrosky and Larry Walker, all of Westinghouse, for their involvement in this issue.

REFERENCES

1   ASME Boiler and Pressure Vessel Code, Section VIII, 1989 Edition

2   ASME Boiler and Pressure Vessel Code, Section III, 1989 Edition

3   Hsu, K. H., Payne, J. R., "PVRC Research Program on Bolted Flanged Connections", Valves, Bolted Joints, Pipe Supports, and Restraints, PVP-Vol. 236, ASME 1992, pp 79-92.

Table 5
Freezing Times for Various Pipe Diameters

	Small		Medium		Large	.
Pipe Sizes (inches)	4	10	12	20	22	30
Freezing Time (minutes)	30	60	100	260	330	640
Average Hold Time (minutes)	60	90	120	180	240	360
Thaw Time (minutes)	15	40	60	120	150	240
Total Time Required (minutes)	105	190	280	560	720	1240
Source: "How to freeze-out downtime on fluid handling systems," Engineer's Digest, June 1983.						

PVP-Vol. 356, Integrity of Structures and Fluid Systems,
Piping and Pipe Supports, and Pumps and Valves
ASME 1997

Use of Motor Power Monitoring in Lieu of Stem Force for Testing MOVs

Brian D. Bunte
Commonwealth Edison Company
Downers Grove, Illinois

ABSTRACT

The purpose of this presentation is to discuss the use of motor power monitoring to perform periodic verification testing of motor operated valves. This presentation discusses the limitations to motor power testing and the benefits which may be derived by using motor power testing in lieu of thrust measurement testing. Examples of test evaluations performed using motor power testing are presented and compared to results obtained when analyzing thrust measurement testing obtained from in situ techniques. The examples and methods discussed in this presentation correspond to rising stem MOVs for which the torque switch is used as the close control switch and the limit switch is used as the open control switch. Methodologies for addressing limit closed MOVs and quarter turn MOVs can be developed along the same lines as those described in this presentation.

PVP-Vol. 356, Integrity of Structures and Fluid Systems,
Piping and Pipe Supports, and Pumps and Valves
ASME 1997

MagIon: Reduce Galling, Control Friction, Save Money: A Technology Assessment

B. R. Black, P.E.
Craig D. Harrington
D. N. Hopkins
Texas Utilities Electric Company
Glen Rose, Texas

ABSTRACT

MagIon® is a vacuum coating process for applying thin metallic films to surfaces. The process can beneficially modify an item's surface characteristics, while not affecting the item's bulk properties. The MagIon process provides substantial margin against galling and effectively controls friction. The Unit 1 reactor vessel head closure studs at TU Electric's Comanche Peak Steam Electric Station (CPSES) nuclear plant have been MagIon treated to eliminate galling-induced stud removal problems. A MagIon surface has also been used to reduce friction at the interface between the valve stem and the packing in a steam generator blowdown valve at CPSES. This presentation reviews technical issues associated with applying the MagIon process and expected benefits.

PVP-Vol. 356, Integrity of Structures and Fluid Systems,
Piping and Pipe Supports, and Pumps and Valves
ASME 1997

DAMAGE AND RESPONSE OF CYLINDRICAL STRUCTURES FROM ASYMMETRIC EXTERNAL AND INTERNAL IMPULSIVE LOADING

Mojtaba Moatamedi, Bruce C. R. Ewan, John L. Wearing

The University of Sheffield, Department of Mechanical and Process Engineering
Mappin Street, Sheffield S1 3JD, UK

ABSTRACT

Nonlinear transient analyses of cylindrical vessels under asymmetric external and internal impulsive loading have been carried out using ANSYS finite element code. The impulsive loading has been chosen to be above the cylinder design pressure, and Von Mises failure criterion has been examined to investigate the failure of the structure. The effects of duration and also three different modelling of impulsive loading, namely triangular, rectangular and exponential, on the response of the structure were investigated. It was observed that by increasing the loading duration, both the maximum displacement, and its occurrence time, increase. A critical curve is introduced to determine the critical impulsive loading and its duration for cylindrical structures. The relations between the transient pressure loading, its duration and the natural frequency of the structure are also explored. It is indicated that the value of impulsive load on cylindrical structures may exceed the design pressure without structural failure.

NOMENCLATURE

E: elastic modulus
E_t: tangent modulus
I: impulse
P: pressure
a: radius
f: frequency
h: thickness
l: length
t: time
θ: angle
ρ: density
ν: Poisson's ratio
σ_y: yield stress

1. INTRODUCTION

The design of cylindrical pressure vessels under impulsive loading is an important subject since cylindrical shell structures have a wide range of applications, some of which are subject to impulsive loading. Although analytical and numerical studies of the dynamics of shells have been carried out for many years, due to the complex kinematics of the response, these have been focused on small displacement and a high degree of symmetry in the geometry and applied loads. As an example, the attempts of Molyneaux et al. (1993) and Pegg (1994) to investigate the failure and response of cylindrical structures subjected to axisymmetric impulsive loading can be mentioned. Lindberg (1964) and Abrahamson and Goodier (1962) have also respectively proposed the elastic and plastic theories for the response of cylindrical shell structures under axisymmetric pulse loading. Some numerical methods such as the analysis discussed by Oslon (1991) have also been introduced for the elasto-plastic response of cylindrical shells under axisymmetric pulse loading since there is no general analytical solution for elasto-plastic behaviour of such structures subjected to pulse loading.

Although those efforts are on the response and failure of cylindrical shell structures under impulsive loading, all attempts have been focused on the axisymmetric loading. Therefore, the damage and response analyses of cylindrical vessel structures under asymmetric external and internal impulsive loading using ANSYS finite element code is investigated here. The results for the external impulsive loading case are compared to existing experimental results (Lindberg and Sliter, 1969) and Von Mises failure criterion was examined for the internal impulsive loading case to investigate the failure of the structure. Three types of explosion modelling, i.e. triangular, rectangular and exponential, are applied to observe the effect of the explosion modelling on the response of the structure. The effects of explosion duration on the response of the structure are also discussed.

2. NUMERICAL ANALYSES

In order to carry out the investigation, numerical analysis has been chosen since implementing the finite element method is able to analyse the structures considering nonlinearities and large deflections. Therefore, a non-linear transient analysis using the ANSYS finite element code which is a general finite element software is used for the present investigation.

2.1. External Impulsive loading

The study of Anderson and Lindberg (1968) on the effects of cylindrical shell parameters on the response of the structures to external pulse loading showed that the behaviour of the cylindrical shell structures under pulse loading strongly depends on the value of the radius to thickness ratio *(a/h)* of the shells so that the elastic behaviour moves to plastic, as the value decreases. Pegg (1991) also obtained similar results and observed that if *(a/h)<60*, the response is plastic, and if *(a/h)>260*, the response is perfectly elastic. For values of *(a/h)* between 60 and 260 the response of the structures to external blast loading is elasto-plastic.

In order to observe the overall response of the cylindrical shells to axisymmetric external impulsive loading, Lindberg and Sliter (1969) carried out an experiment on a cylinder with the following parameters:
Material: Al6061, Length: l=15 cm, Radius: a=7.62 cm, Thickness: h=0.317 cm, Elastic Modulus: E=6.88e10 Pa, Tangent Modulus: E_t=0 Pa, Yield Stress: σ_y= 2.89e8 Pa, Density: ρ=2800 Kg/m³, Poisson's Ratio: ν=0.3.

The ends of the cylinder were taken as clamped. The shell was loaded by a spherical explosive charge off to one side of the shell in their experiment (Lindberg and Sliter, 1969) to which the following empirical formula was fitted to the experimentally measured peak pressures on the rigid cylinder:

$$P(\theta,t) = [(P_R - P_I) \cos^2\theta + P_I].P_0(t) \qquad -90\leq\theta\leq90 \qquad (1)$$
$$P(\theta,t) = P_I P_0(t) \qquad\qquad |\theta|>90 \qquad (2)$$
$$\text{where:} \quad P_0(t) = \exp.[-t/t_0] \text{ and } t_0 = I/P_R \qquad (3)$$

Here, 'I' is the unit impulse measured at θ=0, and P_R and P_I are the reflected and incidence pressure, respectively. The experiment and consequently the finite element analysis undertaken in the paper were carried out for P_R = 5.64e7 Pa, P_I = 8.25e6 Pa and t_0 = 31 μsec.

The response of this cylindrical vessel to explosive blast loading is perfectly plastic, since, according to Pegg (1994) the radius to thickness ratio (a/h) of the cylinder is 24 which is below the limiting value of 60. Therefore, a non-linear analysis for the above cylinder was carried out assuming the stress-strain relationship of the material to be elastic-perfectly plastic.

A modal analysis was first undertaken to determine the natural frequency of the cylindrical vessel, which enables a proper integration time step for the analysis to be calculated. This frequency was determined to be 3265.6 Hz. The transient analysis resulting from exponential loading was then carried out. The

undeformed and permanently deformed shapes of the cylinder are shown in Fig. 1.

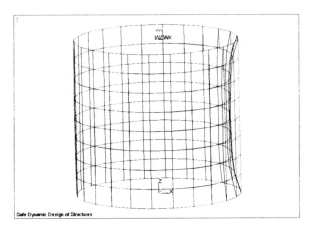

Fig. 1. Undeformed and permanently deformed shapes of the cylinder

Figures 2 and 3, respectively, show the longitudinal and circumferential distributions of the permanent radial displacements from the present finite element analysis as well as the experimental results (Lindberg and Sliter, 1969).

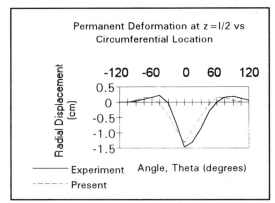

Fig. 2. Permanent circumferential deformation of cylinder

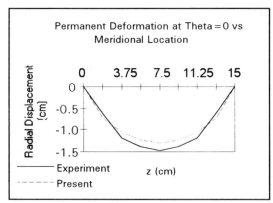

Fig. 3. Permanent meridional deformation of cylinder

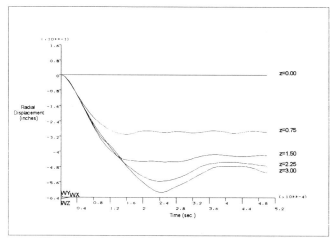

Fig. 4. Transient Meridional Deformation of Cylinder

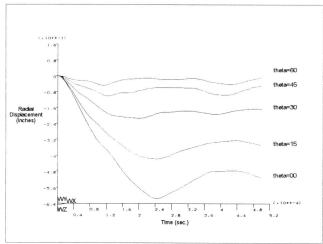

Fig. 5. Transient Circumferential Deformation of Cylinder

A comparison between experimental and numerical results shows some discrepancies which are due to idealised analysis. In other words, the geometry and loading imperfections were not considered in the analysis. In order to obtain closer results to experimental ones, initial imperfections in both shell geometry and loading should be taken into account in the numerical analysis.

However, as can be observed, the results of the present analysis are in good agreement with the experimental ones.

The transient responses (radial displacements), from the present finite element analysis, for $\theta = 0$ along the length of the cylinder, and also for $z = l/2$ with different angles, are illustrated in Figures 4 and 5.

2.2. Internal Impulsive Loading

A 6-metre tube having a diameter of 1 metre made of steel, subjected to internal impulsive spot loading is analysed here. The spot, which has a diameter of 0.5 metre is located in one side on the middle of the tube. The tube was designed for 1 Mpa internal static pressure using ASME boiler and pressure vessel code (ASME, 1983). In order to model the structure in ANSYS, 1512 shell elements with plasticity, large deflection, and large strain capabilities are used as shown in Figure 6. Both ends of the cylinder are assumed to be clamped.

Rectangular type of impulsive loading is used for all analyses so that the load suddenly increases to the maximum value and then remains for the limited duration and finally returns to zero.

The tube under impulsive loading is expected to have large deformations and thus large displacements and rotations which were taken into account by invoking the nonlinear strain-displacement relationships where higher-order derivatives of displacements and rotations are included. Nonlinearity of the material is assumed to follow the bilinear isotropic hardening. This option uses the Von Mises yield criterion to assess the failure of the structure under the specified loading conditions. Damping is also taken into account to increase the accuracy of the final deflection of the structure in order to obtain the realistic results.

The modal analysis of the tube was first undertaken to determine the natural frequency of the structure which obtained equal to 66.7 Hz (Period: 15 msec). A series of analyses were then carried out to observe the response of the structure to internal impulsive spot loading with the value of 4 Mpa and 10-50 msec duration. The failure of the structure, for each case, was checked by applying Von Mises criterion to determine the failure point of the structure.

The transient response of the centre of the spot obtained from the nonlinear transient analysis with 4 Mpa pressure and 10 msec duration, is shown in Fig. 7.

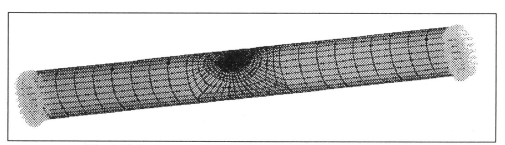

Fig. 6. Finite element model of the circular cylindrical shell structure

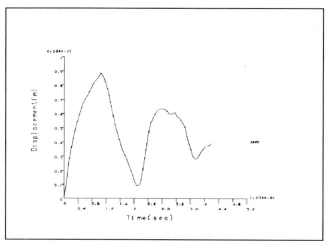

Fig. 7. Displacement-time relationship for the centre of the spot

$$P_0(t) = \exp.\ [-t/t_0]\ \&\ t_0 = 31\ \mu sec. \qquad (4)$$

$$I = \int P.dt \qquad (5)$$

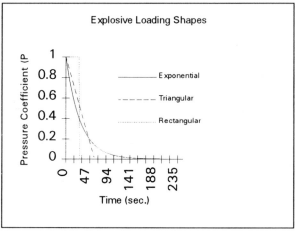

Fig.8. Impulsive loading shapes

As can be observed, the displacement of the point increases to a maximum value and then damps to a final deflection which is not zero. A combination of the structural modes are excited in the dynamic response and therefore the natural frequency of the structure plays an important role in the structural response. If the pulse duration is very short compared to the structural response (the structural response is defined as the period of the structure which equals '1/f'), the response of the structure is elastic but a long duration results in a plastic response. For the present analysis, the duration is very close to the period of the structure so that the response is consequently elasto-plastic. The above explanation is correct when the pulse amplitude is kept constant.

The maximum stress in the structure reaches a value which is higher than the yield stress, however the structure does not fail under this condition. The reason is that the pulse duration is too short to cause the failure so that the load is not categorised in static loading. It indicates that the structure can endure stresses higher than yield stress for less than a critical duration, in other words the dynamic yield criteria are different form the static ones.

Von Mises failure criterion was applied to assess the structural failure for the present case. However the nonlinear transient analyses of the shell structure under internal transient spot loading with different values for pressure and duration can produce a critical curve which indicates a safe region for tubes under impulsive spot loading without structural failure. The generation of the critical curve for the present case is experimentally under investigation by the authors.

2.3. Explosion Modelling And Duration

For the external explosion case, the explosive loading was modelled as an exponential curve. Triangular and rectangular modelling are now considered, and a comparison of the effects of these three types of loading, i.e. triangular, rectangular and exponential, on the response of the cylindrical vessel to external explosion is made. In order to achieve this purpose, $P_0(t)$ needs to be changed in equations (1) and (2), so that 'I', which is the total impulse, remains equal for all cases. In other words:

These three shapes are illustrated in Fig.8.

The above three shapes of external explosive loading were applied to the cylindrical vessel and results for $\theta = 0$ and $z = l/2 = 3$ in. are illustrated in Fig. 9.

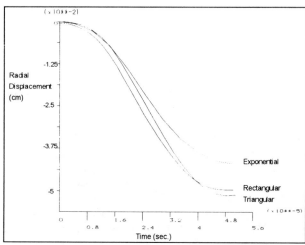

Fig. 9. Transient deformation for different modelling

A comparison of the obtained results reveals that the triangular and rectangular modelling give a greater value of the maximum displacement of the structure than the exponential modelling. This comparison of the effects of pulse shape on the response of the structures can be very significant in practical work where measurement and control of the shape of explosive loading can be very difficult.

Following the analyses of the cylindrical structure under external explosion, the effects of the duration of the explosion are now investigated by changing t_0 in equation (4) for exponential modelling.

The maximum displacement of the shell occurs in the position with $\theta=0$ & $z=l/2=3$ in., which is the critical point of the analyses. Therefore, the transient analyses of this point for different values of t_0 were carried out and it was observed that by increasing the duration of the explosion, both the maximum displacement, and the time in which the maximum displacement occurs, increase.

3. FUTURE WORK

Figure 10 represents a simplified diagram of the experiment which are being used to investigate the failure and response of the structures under asymmetric impulsive loading.

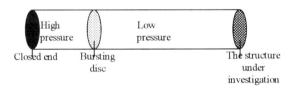

Fig. 10. Experimental Schematic

The bursting disc is ruptured at a certain pressure and consequently a shock wave moves towards the structure. The final pressure can not be atmospheric pressure as the above arrangement is in the confined vessel. Some arrangements have been made to reduce the pressure which may be achieved by providing some holes around the structure which leads to exit of the high pressure gas to the atmosphere. A large shock tube is also available to investigate the response of large structures to explosions.

4. CONCLUSIONS

The transient analysis of a cylindrical vessel under asymmetric external impulsive loading was carried out numerically using the finite element method. The results were compared with existing experimental results and good agreements were obtained. The failure of a cylindrical structure due to an internal impulsive spot loading was also investigated numerically. The response of the cylindrical vessel was found to be elasto-plastic since the pulse duration was close to the period of the structure. Von Mises failure criterion were applied to investigate the failure of the structure. According to the analysis carried out by ANSYS code, it was found that a dynamic pressure greater than the design pressure can be applied to the structure without failure if the impulse duration is less than a critical value. Three different types of modelling for the external explosion, i.e. triangular, rectangular and exponential, were considered. It was found that the exponential modelling agrees well with existing experimental results and is more realistic than the other two types of modelling considered in the paper. The

effects of explosion duration on the response of the structure under external explosion in the plastic response region were also investigated. It was found that by increasing the duration of the explosion, both the maximum displacement, and the time in which the maximum displacement occurs, increase. It can be concluded that the design criteria for structures under dynamic loading are more flexible since the dynamic yield criteria are different from static ones.

5. REFERENCES

Abrahamson, G.R. and Goodier, J.N., 1962, "Dynamic plastic flow buckling of a cylindrical shell from uniform radial impulse", *Forth US national congress of applied mechanics 2*

American Society of Mechanical Engineers (ASME), 1983, "Article D-3, Shells of revolution under external pressure", *Section VIII, Pressures vessels, Division 2, Alternative rules*, 225-230

Anderson, D. L.; Lindberg, H. E., 1968, "Dynamic Pulse Buckling of Cylindrical Shells Under Transient Lateral Pressures", AIAA Journal, Volume 6, Number 4, 589-598

Lindberg, H. E. and Sliter, G. E., 1969, "Response of Reentry-Vehicle-Type Shells to Transient Surface Pressures", AFWL-TR-68-56, Report by SRI International to Air Force Weapons Laboratory

Lindberg, H.E, 1964, "Buckling of a very thin cylindrical shell due to an explosive pressure", J Applied Mech, 31: 261-272

Molyneaux, T.C.K., Li, Long-Yuan and Firth, N., 1993, "Impact response of circular cylindrical shells under explosive loading", Adv Eng Soft, 18: 7-13

Oslon, M.D., 1991, "Efficient modelling of blast loaded plate and cylindrical shell structures", Comp & struc, 40: 1139-1149

Pegg, N. G., 1991, "Dynamic Pulse Buckling of Cylinders of Various a/h Ratios", Computers and Structures, Volume 39, Number 1/2, 173-183

Pegg, N.G., 1994, "A numerical study of dynamic pulse buckling of cylindrical structures", Marine struc, 7: 189-212

PVP-Vol. 356, Integrity of Structures and Fluid Systems,
Piping and Pipe Supports, and Pumps and Valves
ASME 1997

EFFECT OF GEOMETRIC GAP BETWEEN CYLINDER AND REINFORCEMENT PAD ON LOCAL STRESSES (MOMENT LOADING ON NOZZLE)

Z.F. Sang, L. Li and Y.J. Zhou
Department of Mechanical Engineering
Nanjing University of Chemical Technology
Nanjing, P.R. China

G.E.O. Widera
Department of Mechanical and Industrial Engineering
Marquette University
Milwaukee, WI 53201 USA

ABSTRACT

The purpose of this paper is to present a study of the effect of a geometric gap between the cylindrical shell and reinforcement pad on the local stresses in the area of the intersection when subjected to nozzle moments. Experimental and finite element analyses were performed on two test vessels (four nozzles). A comparative study of stresses in the intersection region for different geometric gaps was also carried out.

KEYWORDS

Opening-reinforcement, moments on nozzle, local stress, electrical resistance strain gage measurement, finite element analysis.

NOMENCLATURE

D_i	inside diameter of cylinder
D	average diameter of cylinder
D_1	outer diameter of pad
T	thickness of cylinder
T_1	thickness of pad
L	length of cylinder
d_i, d_o	inside, outside diameter of nozzle($d_i = 100$ mm, $d_o = 108$ mm)
d	nominal diameter of nozzle
r	nominal radius of nozzle
t	thickness of nozzle
l	axial length of nozzle
L_1	half length of cylinder
θ	angle around nozzle
X	distance from the intersection along cylinder
Y	distance from the intersection along nozzle
M_L	longitudinal moment on nozzle
M_C	transverse moment on nozzle
α_k	impact toughness
δ_s	percentage of elongation
σ	maximum stress
σ_b	ultimate strength
σ_s	yield strength
$\sigma_{i\theta}$	circumferential stress of inner surface
σ_{ix}	longitudinal stress of inner surface
$\sigma_{o\theta}$	circumferential stress of outer surface
σ_{ox}	longitudinal stress of outer surface

INTRODUCTION

A reinforcement structure consisting of a pad is an important type of local reinforcement in pressure vessels and piping connections because of its simple structure, convenient manufacture, low cost and rich application experience. As a result, it is widely used for applications with low or medium pressure, especially those with small fluctuations in pressure and temperature. For the design of the reinforcement pad, the current design code (ASME, 1989) states a specific rule. However, only the design method is given; it does not provide the technique for calculating the stresses in the reinforcement region. As a result the distribution and the magnitudes of the local stresses induced by the geometric discontinuity and the load due to reinforcement is not known. As for loads, the design code only considers the effect of internal pressure. But in fact, such loadings as axial thrust, moments, twist and shear, etc. on the nozzle are usually applied to pressure vessels, process equipment or piping connections. Besides, designers require that manufacturers keep a perfect contact between the reinforcement pad and cylinder during the fabrication of the vessel to ensure that it has the ability to carry the various loadings as required. But, for a variety of different reasons, perfect contact cannot be kept between cylinder and pad and, thus, a gap results. The effect of the gap on the stresses in the nozzle-reinforcement region is a matter of common interest to both designers and manufacturers.

The stress analysis of the reinforcement pad structure represents a complicated problem. Research is mostly focused on the analysis of spherical shells with radial nozzles. Soliman and Gill (1978) and Oikawa and Oka (1987) carried out analyses and research on the reinforcement pad structure of such shells by both theoretical analysis and FEM technology. Other researchers (Chao et al., 1986) also presented valuable research results in the design and analysis of the reinforcement pad structure. WRC 107 (1979) and WRC 297 (1984) are the two most authoritative design documents for analyzing and calculating the local stresses in the opening-nozzle region under the action of external loads, and are widely used. However, they do not present specific design methods when there exists a reinforcement structure at the nozzle intersection. This paper presents a detailed study of the effect of a geometric gap between the cylindrical shell with radial nozzle and the reinforcement pad on the local stresses in the area of the nozzle-cylinder-pad intersection under the action of nozzle moments (including longitudinal and transverse moments) by both an experimental study as well as a finite element analysis. A comparative study of the stress distribution and numerical results of the stresses in the intersection region with different geometric gaps is also carried out.

EXPERIMENTAL DETAILS

Test Vessels

Two specially designed and manufactured test vessels (No. 1 and No. 2) were used in the experiments. The configuration of vessel No. 2 is shown in Fig. 1. Each vessel has two 180° symmetrically installed radial nozzles of the same size. One nozzle of vessel No. 1 was not reinforced, while the other nozzle was reinforced with a standard reinforcement pad; there was no geometric gap between the cylindrical shell and reinforcement pad. The two nozzles of vessel No. 2 were reinforced by reinforcement pads of the same size. The geometric gaps between the cylindrical shell and reinforcement pads were 2 and 4 mm, respectively. The specific structural dimensions of the two test vessels are listed in Table 1 and shown in Fig. 2.

TABLE 1 STRUCTURAL DIMENSIONS OF THE TEST VESSELS

Vessel No.	D mm	L mm	L_1 mm	T mm	D/T	Nozzle No.	d mm	t mm	l mm	T_1 mm	D_1 mm	d/D	l/T	Gap δ mm
No. 1						3	100	4	170			0.2	0.5	Without pad
	500	800	400	8	62.5	4	100	4	170	8	200	0.2	0.5	No
No. 2						7	100	4	170	8	200	0.2	0.5	2
						8	100	4	170	8	200	0.2	0.5	4

Chemical Composition and Properties of the Materials Used in the Test Vessels

The materials used for each component of the two test vessels are the same. These materials, their chemical composition and mechanical properties are summarized in Table 2.

TABLE 2 MATERIALS AND PROPERTIES FOR THE TEST VESSELS

Name of Parts	Material	Chemical Composition %					Tension Test			Impact Test		
		C	Si	Mn	P	S	σ_b Mpa	σ_s Mpa	δ_s %	Notch Type	Test Temp.	α_k J/cm²
Head Cylinder Pad	16MnR	0.19	0.39	1.48	0.027	0.020	605	435	27.0	V	Ambient	22 24 26
Nozzle	20	0.19	0.27	0.46	0.019	0.017	436	267	34.0	V	Ambient	100 146 154

Local Structure of Reinforcement Region

Extruding nozzle structures with inside and outside fillet welds were used for the four nozzles of the two test vessels. The reinforcement pads were located on the outside of the cylinders. The local structures and gaps in the reinforcement regions are shown in Fig 2. In order to keep the gaps uniform during fabrication, four spacers of 2 mm and 4 mm thickness were put under the pads and arranged evenly around the circumference of the pad; the pads were then welded.

Test Method and Procedure

The electrical resistance strain gage measurement method was used during the test phase of the study. The strain gages were installed in the longitudinal (θ = 0° ~180°) and transverse (θ = 90° ~270°) sections of the cylinders and nozzles. The gages at each measuring point consisted of 90° biaxial gages. The locations of the strain gages in the nozzle-reinforcement region of vessel No. 2 are shown in Figure 3. The loads were applied to the ends of the nozzles by a 30T separate-type hydraulic jack. The structure of the loading set-up and method are illustrated in Figure 4. The test procedure, and the instruments and data recording employed are summarized in Sang et al. (1995).

Test Results

The test results indicate that no matter what geometric gap exists between the reinforcement pad and the cylindrical shell, the stresses in the opening-reinforcement pad region, due to moments on the nozzle, possess the following behavior: the stress in the transverse (θ = 90° ~270°) section of the cylinder under longitudinal moment M_L on nozzle is very small, it can be ignored; the maximum stress occurs in the longitudinal (θ = 0° ~180°) section. Similarly, the stress in the longitudinal (θ = 0° ~180°) section of cylinder under transverse moment M_C on the nozzle is very small, it can also be ignored; the maximum stress occurs in the transverse (θ = 90° ~270°) section. The distributions of stresses in the compressed sides in the longitudinal (θ = 180°) and transverse (θ = 90°) sections of cylinder and nozzle due to the longitudinal moment M_L = 6.27 KN-m and transverse moment M_C = 3.14 KN-m on the nozzle for the four nozzles of the two test vessels are shown in Figures 5 to 8, respectively. The abscissas X, Y in the figures are the axial distances from any point on the cylinder or nozzle to the point of cylinder-nozzle intersection. The ordinate is the value of the test stress.

FINITE ELEMENT ANALYSIS

Because of the purpose of this paper is to determine the local stresses in the opening-reinforcement region, a three dimensional 8-node brick element has been used for the finite element analysis with one element through the thickness. Due to the non-symmetry of the structure when the nozzle is subjected to the transverse moment, the whole vessel is taken in setting up the calculation model. However, when the nozzle is subjected to the longitudinal moment, only one-half vessel is taken. Figure 9 indicates the FEM mesh (total of 596 elements). The loads are applied uniformly at every node of the ends of the nozzles along the longitudinal and transverse directions of cylinder. Because the bottom of the test vessels was rigidly clamped by bolts to the base plate, it is regarded as a fixed boundary in the FEM analyses. It is worth pointing out that both the outside and inside fillet welds of the nozzle and cylinder, and the fillet welds of the cylinder and reinforcement pad are taken into account in these FEM analyses. It is thus seen that the analysis models are not simplified ones but are consistent with the actual test models.

The deformation in the nozzle-reinforcement region of nozzle No. 4 under longitudinal moment M_L = 6.27 KN-m and transverse moment M_C = 3.14 KN-m are indicated in Figures 10.a and 10.b, respectively. Figures 11.a and 11.b illustrate the maximum principle stress variation in the compressed sides (θ = 180°, θ = 90°) of the opening-reinforcement region induced by a longitudinal moment M_L = 6.27 KN-m and transverse moment M_C = 3.14 KN-m for nozzle No. 4 (geometric gap between the cylindrical shell and reinforcement pads is 4 mm), respectively. Figures 11.a and 11.b indicate that the stresses in the cylinders and nozzles induced by moments on the nozzle possess an obvious localized behavior.

COMPARISONS AND CONCLUSIONS

1.   Table 3 shows the comparisons between the test and FEA results. The stress ratio in this table is the maximum stress in the cylinders or nozzles divided by the nominal stress induced by the moments on the nozzles.

    The stresses in Table 3 are the measured values, they are not extrapolated to the junction, as a result they are lower than those from the FEM analysis. From Table 3, one can see that the maximum stress is located in the transverse section of the cylinder and outside of the cylinder-pad weld under the action of the same moments on the nozzles. It is caused by the higher circumferential bending induced by the stiffening effects of the reinforcement pad and welds and the discontinuity of the geometric shapes.

2.   Both the test and FEA results indicate that the stiffening effect of the reinforcement pad is obvious whether there is a geometric gap between the cylindrical shell and reinforcement pad or not. It greatly reduces the stress concentration at the edge of the opening.

3.   The test and FEA results indicate that the stresses in the cylinders and nozzles possess an obvious localized behavior, no matter what the numerical values of the geometric gaps are. The attenuation cycle of stress in the nozzle is about 30 mm, which is $2.1\sqrt{rt}$. That in the cylinder is about 120 mm, which is $2.7\sqrt{RT}$.

4.   The stresses in the compression sides (θ = 180°, θ = 90°) and the tension sides (θ = 0°, θ = 270°) under moments on the nozzle are obviously antisymmetric about the central line of the nozzle no matter what the numerical values of the geometric gaps between the cylindrical shells and reinforcement pads are. That is to say, the absolute values of the stresses are the same, the distributions are the same, but the signs of the stresses are opposite. It is for this reason that this paper only presents the stresses in the compression sides.

5.   The stresses in the transverse section of the cylinders (or reinforcement pads) induced by transverse moment M_C are much larger than the stresses in the longitudinal section of the cylinders (or reinforcement pads) induced by longitudinal moment M_L. The stress ratio of these two sections induced by a unit moment on the nozzles are listed in Table 4.

TABLE 4 TEST RESULTS OF THE TRANSVERSE-LONGITUDINAL STRESS RATIO

Vessel No.	Nozzle No.	Geometric Gap δ mm	Transverse-longitudinal Stress Ratio $(\sigma / M_C) / (\sigma / M_L)$
No. 1	No. 1	Without pad	1.79
	No. 2	δ = 0	2.27
No. 2	No. 3	δ = 2	2.51
	No. 4	δ = 4	2.31

6.   For a loading of moments on nozzles, the effect of the geometric gap between the reinforcement pad and cylinder on the local stresses in the reinforcement region is not obvious. That is to say that the stresses in the cylinders and nozzles with a certain size geometric gap between the reinforcement pad and cylindrical shell are not obviously larger than those without a geometric gap.

REFERENCES

ASME Boiler and Pressure Vessel Code, 1989, Section 8, Division 1, American Society of Mechanical Engineers, New York.

Chao, Y.J., Wu, B.C., and Sutton, M.A., 1986, "Radial Flexibility of Welded-Pad Reinforced Nozzles in Ellipsoidal Pressure Vessel Heads," *Int. J. Pres. Ves. And Piping*, Vol. 24, pp. 189-207.

TABLE 3 COMPARISONS OF RESULTS

Vessel No.	Nozzle No.	Geometric Gap δ mm	Section	Moments on Nozzle KN-m	Maximum Stress Mpa Test Cylinder	Maximum Stress Mpa Test Nozzle	Maximum Stress Mpa FEM	Stress Ratio* Test Cylinder	Stress Ratio* Test Nozzle	Stress Ratio* FEM
No. 1	No. 1	Without pad	θ=180	M_L=627KN-m	-347	-260	-466 (Cylinder)	1.82	1.36	2.44
			θ=90°	M_C=3.14KN-m	-311	-250	-350 (Cylinder)	3.27	2.63	3.68
	No. 2	δ = 0 With pad	θ=180	M_L=6.27KN-m	-148	-200	-282 (Nozzle)	0.77	1.05	1.48
			θ=90°	M_C=3.14KN-m	-165	-155	-160 (Cylinder)	1.79	1.63	1.68
No. 2	No. 3	δ = 2	θ=180	M_L=6.27KN-m	-135	-210	-247 (Nozzle)	0.68	1.10	1.93
			θ=90°	M_C=3.14KN-m	-170	-135	-155 (Cylinder)	1.84	1.42	1.63
	No. 4	δ = 4	θ=180	M_L=6.27KN-m	-130	-230	-274 (Nozzle)	0.65	1.21	1.43
			θ=90°	M_C=3.14KN-m	-150	-125	-132 (Cylinder)	1.63	1.32	1.39

*Stress ratio = $\sigma / (M_{L(C)} / Z_b)$

Where:
σ maximum stress of experiment or calculation, MPa;
$M_{L(C)}$ longitudinal (transverse) moment on nozzles, KN-m;
Z_b section modules of bending, mm^3.

$$Z_b = \frac{\pi(d_o^2 - d_i^4)}{32 d_o}$$

Mershon, J.L., Mokhtarian, K., Ranjan, G.V., and Rodabaugh, E.C., 1984, "Local Stresses in Cylindrical Shells Due to External Loading on Nozzle-Supplement to WRC Bulletin No. 107," *WRC Bulletin No. 297*.

Oikawa, T. And Oka, T., 1987, "A New Technique for Approximating the Stress in Pad-Type Nozzles Attached to a Spherical Shell," *Journal of Pressure Vessel Tech.*, Vol. 109, pp. 188-192.

Sang, Z.F., Li, L., Qian, H.L., and Widera, G.E.O., 1995, "Behavior of Pad Reinforced Cylindrical Vessels Subjected to Axial Thrust on Nozzle," PVP-Vol. 318, the *1995 Joint ASME/JSME Pressure Vessels and Piping Conference*, p. 1.

Soliman, S.F. and Gill, S.S., 1978, "Radial Loads on Pad-Reinforced Nozzles in Spherical Pressure Vessels - A Theoretical Analysis and Experimental Investigation," *Int. J. Pres. Ves. And Piping*, Vol. 6, pp. 451-472.

Wichman, K.R., Hopper, A.G., and Mershon, J.L., 1979, "Local Stresses in Spherical and Cylindrical Shells Due to External Loading," *WRC Bulletin No. 107*.

Figure 1   Configuration of test vessel No.2

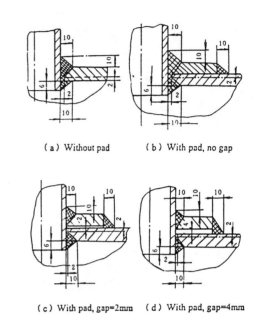

(a) Without pad   (b) With pad, no gap

(c) With pad, gap=2mm   (d) With pad, gap=4mm

Figure 2   Local structure in the nozzle-cylinder-pad region

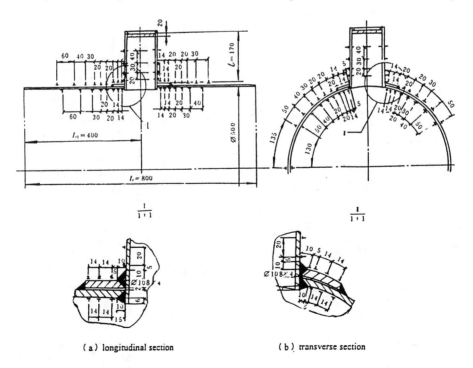

(a) longitudinal section   (b) transverse section

Figure 3   Location of the strain gages of the vessel No.2

154

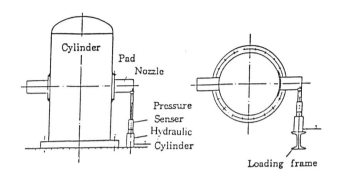

longitudinal moment M_L Transverse moment M_C

Figure 4   Loading method

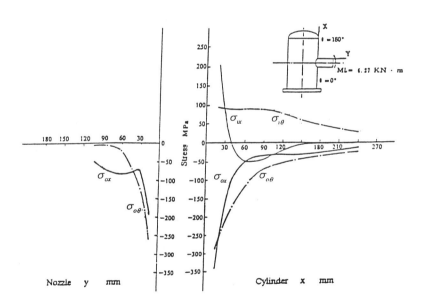

Figure 5.a   Stresses of longitudinal section induced by M_L (Nozzle No.3, $\theta = 180°$)

155

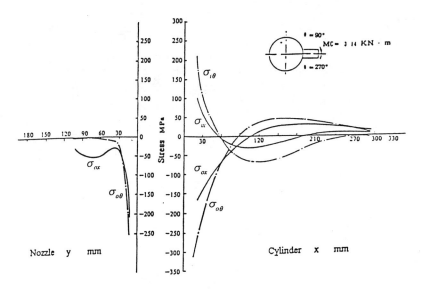

Figure 5.b   Stresses of transverse section indused by M_C (Nozzle No.3, $\theta = 90°$)

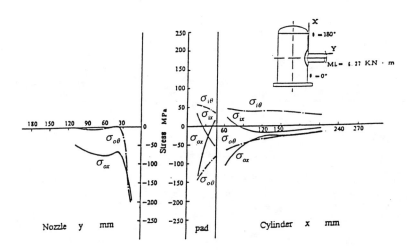

Figure 6.a   Stresses of longitudinal section induced by M_L (Nozzle No.4, $\theta = 180°$)

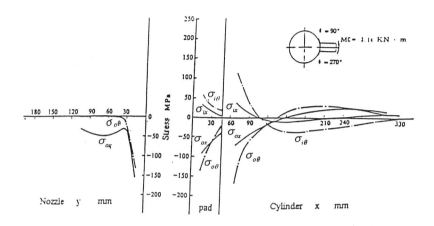

Figure 6.b   Stresses of transverse section indused by M_C (Nozzle No.4, $\theta = 90°$)

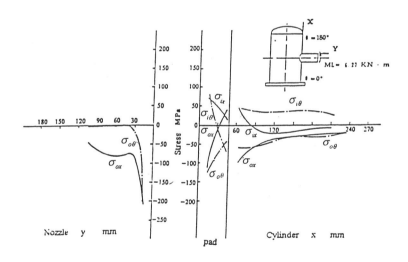

Figure 7.a   Stresses of longitudinal section induced by M_L (Nozzle No.7, $\theta = 180°$)

157

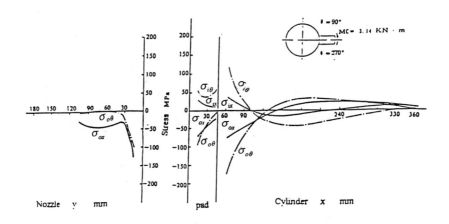

Figure 7.b   Stresses of transverse section indused by M_C (Nozzle No.7, $\theta = 90°$)

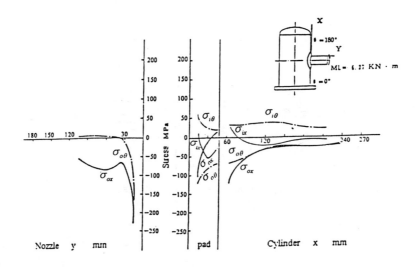

Figure 8.a   Stresses of longitudinal section induced by M_L (Nozzle No.8, $\theta = 180°$)

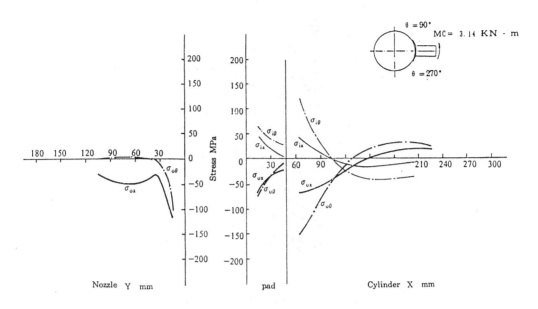

Figure 8.b   Stresses of transverse section induced by M_C (Nozzle No.4, $\theta = 90°$)

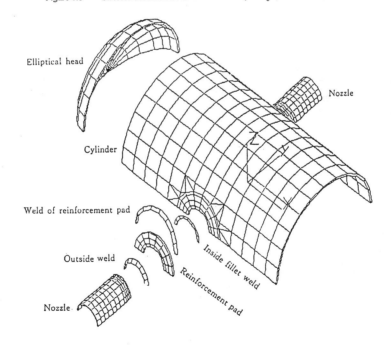

Figure 9   Meshes of FEM analysis

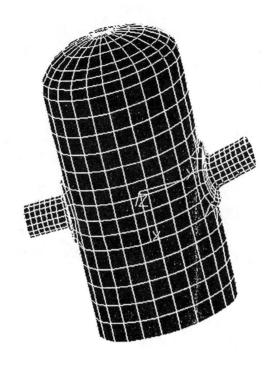

Figure 10.a   Longitudinal deformation

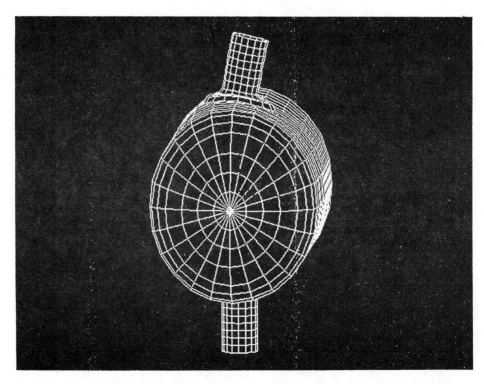

Figure 10.b   Transverse deformation

160

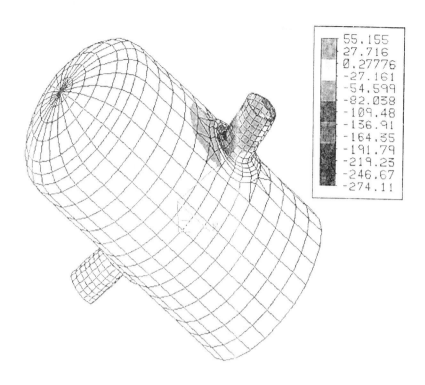

Figure 11.a   Stresses induced by M_L (Nozzle No.4, $\theta = 180°$)

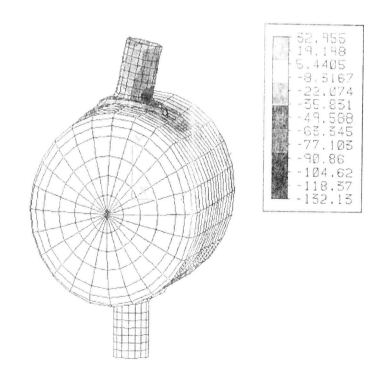

Figure 11.b   Stresses induced by M_c (Nozzle No.4, $\theta = 90°$)

161

AUTHOR INDEX

PVP-Vol. 356
**Integrity of Structures and Fluid Systems, Piping and Pipe Supports,
and Pumps and Valves — 1997**